Aha! Solutions

© 2009 by
The Mathematical Association of America (Incorporated)
Library of Congress Catalog Card Number 2008938665
ISBN: 978-0-88385-829-5
Printed in the United States of America
Current Printing (last digit):
10 9 8 7 6 5 4 3 2 1

Aha! Solutions

Martin Erickson
Truman State University

Published and Distributed by
Mathematical Association of America

Council on Publications
Paul Zorn, *Chair*

Problem Books Editorial Board
Richard A. Gillman *Editor*
Roger Nelsen
Mark Saul
Tatiana Shubin

MAA PROBLEM BOOKS SERIES

Problem Books is a series of the Mathematical Association of America consisting of collections of problems and solutions from annual mathematical competitions; compilations of problems (including unsolved problems) specific to particular branches of mathematics; books on the art and practice of problem solving, etc.

Aha! Solutions, Martin Erickson

The Contest Problem Book VII: American Mathematics Competitions, 1995–2000 Contests, compiled and augmented by Harold B. Reiter

The Contest Problem Book VIII: American Mathematics Competitions (AMC 10), 2000–2007, compiled and edited by J. Douglas Faires & David Wells

The Contest Problem Book IX: American Mathematics Competitions (AMC 12), 2000–2007, compiled and edited by David Wells & J. Douglas Faires

A Friendly Mathematics Competition: 35 Years of Teamwork in Indiana, edited by Rick Gillman

The Inquisitive Problem Solver, Paul Vaderlind, Richard K. Guy, and Loren C. Larson

International Mathematical Olympiads 1986–1999, Marcin E. Kuczma

Mathematical Olympiads 1998–1999: Problems and Solutions From Around the World, edited by Titu Andreescu and Zuming Feng

Mathematical Olympiads 1999–2000: Problems and Solutions From Around the World, edited by Titu Andreescu and Zuming Feng

Mathematical Olympiads 2000–2001: Problems and Solutions From Around the World, edited by Titu Andreescu, Zuming Feng, and George Lee, Jr.

Problems from Murray Klamkin: The Canadian Collection, edited by Andy Liu and Bruce Shawyer

The William Lowell Putnam Mathematical Competition Problems and Solutions: 1938–1964, A. M. Gleason, R. E. Greenwood, L. M. Kelly

The William Lowell Putnam Mathematical Competition Problems and Solutions: 1965–1984, Gerald L. Alexanderson, Leonard F. Klosinski, and Loren C. Larson

The William Lowell Putnam Mathematical Competition 1985–2000: Problems, Solutions, and Commentary, Kiran S. Kedlaya, Bjorn Poonen, Ravi Vakil

USA and International Mathematical Olympiads 2000, edited by Titu Andreescu and Zuming Feng

USA and International Mathematical Olympiads 2001, edited by Titu Andreescu and Zuming Feng

USA and International Mathematical Olympiads 2002, edited by Titu Andreescu and Zuming Feng

USA and International Mathematical Olympiads 2003, edited by Titu Andreescu and Zuming Feng

USA and International Mathematical Olympiads 2004, edited by Titu Andreescu, Zuming Feng, and Po-Shen Loh

MAA Service Center
P. O. Box 91112
Washington, DC 20090-1112
1-800-331-1622   fax: 1-301-206-9789

*To Martin Gardner and Ross Honsberger
whose mathematical exposition inspires so many*

Preface

Every mathematician (beginner, amateur, and professional alike) thrills to find simple, elegant solutions to seemingly difficult problems. Such happy resolutions are called "aha! solutions," a phrase popularized by mathematics and science writer Martin Gardner in his books *Aha! Gotcha* and *Aha! Insight*. Aha! solutions are surprising, stunning, and scintillating: they reveal the beauty of mathematics.

This book is a collection of problems whose aha! solutions I enjoy and hope you will enjoy too. The problems are at the level of the college mathematics student, but there should be something of interest for the high school student, the teacher of mathematics, the "math fan," and anyone else who loves mathematical challenges.

As a student first learning mathematics, I was inspired by the works of Martin Gardner and mathematics expositor Ross Honsberger (I still am today). One of the best ways to capture the imagination of young people and get them interested in mathematics is by "hooking them" on irresistible problems. And such a hook is appropriate, since a great part of mathematical study and investigation consists of problem solving. Sometimes the problem solving is at an advanced level, sometimes it is what we discover and create in our everyday mathematical lives.

For this collection, I have selected one hundred problems in the areas of arithmetic, geometry, algebra, calculus, probability, number theory, and combinatorics. Some of the problems I created, others are old but deserve to be better known. The problems start out easy and generally get more difficult as you progress through the book. A few solutions require the use of a computer. An important feature of the book is the bonus discussion of related mathematics that follows the solution of each problem. This material is there to entertain and inform you or point you to new questions. If you don't remember a mathematical definition or concept, there is a Toolkit in the back of the book that will help.

I take to heart the poet Horace's decree that writing should delight and instruct. So I hope that you have fun with these problems and learn some new mathematics. Perhaps you will have the satisfaction of discovering aha! solutions of your own.

Thanks to the following people who have provided suggestions for this book: Robert Cacioppo, Robert Dobrow, Christine Erickson, Suren Fernando, Martin Gardner, David

Garth, Joe Hemmeter, Ross Honsberger, Daniel Jordan, Ken Price, Khang Tran, and Anthony Vazzana.

Contents

Preface vii

1 **Elementary Problems** 1
   1.1 Arithmetic 1
        Fair Division 1
        A Mere Fraction 2
        A Long Sum 4
        Sums of Consecutive Integers 5
        Pluses and Minuses 7
        Which is Greater? 8
        Cut Down to Size 10
        Ordered Digits 11
        What's the Next Term? 12
   1.2 Algebra 13
        How Does She Know? 13
        How Cold Was It? 14
        Man vs. Train 15
        Uphill, Downhill 17
        How Many Solutions? 18
   1.3 Geometry 19
        A Quadrilateral from a Quadrilateral 19
        The Pythagorean Theorem 20
        Building Blocks 22
        A Geometric Inequality 23
        A Packing Problem 25
        What's the Area? 26
        Volume of a Tetrahedron 27
        Irrational ϕ 28
        Tangent of a Sum 29

	1.4	No Calculus Needed	32
		A Zigzag Path	32
		A Stack of Circles	33
		A Farmer's Field	35
		Composting—A Hot Topic?	36
		Three Sines	37

2 Intermediate Problems   41

	2.1	Algebra	41
		Passing Time	41
		Sums to 1,000,000	43
		An Odd Determinant	44
		Demanding a Polynomial	46
	2.2	Geometry	46
		What's the Side Length?	46
		Napoleon's Theorem	49
		A Graph on a Doughnut	51
		Points Around an Ellipse	53
		Three Fixed Points	54
		'Round and 'Round	56
		Reflections and Rotations	58
		Cutting and Pasting Triangles	60
		Cookie Cutting	62
		Revolving Credit	63
	2.3	Calculus	67
		The Harmonic Series	67
		A Quick Integral	68
		Euler's Sum	69
		Strips of Carpeting	70
		Is it 0?	71
		Pi is Pi	72
	2.4	Probability	74
		How Many Birthdays?	74
		The Average Number of Spots	75
		Balls Left in an Urn	76
		Random Points on a Circle	80
		The Gobbling Algorithm	80
	2.5	Number Theory	83
		Square Triangular Numbers	83
		Foxy Factorial	84
		Always Composite	85
		A Problem of 1's	86
		A Problem of 2's	87
		A Problem of 3's	88
		Fibonacci Squared	90

		A Delight from Pascal's Triangle	93
		An Unobvious Integer	93
		Magic Squares	96
		All Things Being Equal	99
	2.6	Combinatorics	100
		Now I Know My ABC's	100
		Packing Animals in a Box	101
		Linear Bumper Cars	103
		I Scream Aha!	106
		Lines Dividing the Plane	107
		A Number that Counts	108
		A Broken Odometer	109
		How Many Matrices?	110
		Lots of Permutations	111
		Even Steven and Oddball	113
		Higher-Dimensional Tic-Tac-Toe	115
		The Spice of Life	117
		Sperner's Lemma	120
		An Infinite Series	122
		Change for a Dollar	123
		Rook Paths	124
3	Advanced Problems		129
	3.1	Geometry	129
		Self-Intersecting Polygons	129
		Regular Simplices	131
		$n^2 + 1$ Closed Intervals	134
	3.2	Probability	135
		1,000,000 Coin Flips	135
		Bits of Luck	138
		A Game for Noncommunicating Mathematicians	140
	3.3	Algebra	143
		An Integer Matrix with Determinant 1	143
		Only $1, -1, 0$?	146
		168 Elements	148
	3.4	Number Theory	154
		Odd Binomial Coefficients	154
		Fibonacci Factors	156
		Exact Covering Systems	158
		A Fibonacci Number Producing Polynomial	160
		Perrin's Sequence	162
	3.5	Combinatorics	165
		Integer Triangles	165
		Topsy-Turvy Tournaments	167
		Sudoku Solving	169

	The SET Game	172
	Girth Five Graphs	175
	Unlabeled Graphs	178

A Toolkit 183

B List of Bonuses 193

Bibliography 197

Index 199

About the Author 207

1
Elementary Problems

Let's begin with some relatively easy problems. The challenges become gradually more difficult as you go through the book. The problems in this chapter can be solved without advanced mathematics. Knowledge of basic arithmetic, algebra, and geometry will be helpful, as well as your own creative thinking. I recommend that you attempt all the problems, even if you already know the answers, because you may discover new and interesting aspects of the solutions. A bonus after each solution discusses a related mathematical topic. Remember, each problem has an aha! solution.

1.1 Arithmetic

Fair Division

Abby has fifteen cookies and Betty has nine cookies. Carly, who has no cookies, pays Abby and Betty 24 cents to share their cookies. Each girl eats one-third of the cookies.

Betty says that she and Abby should divide the 24 cents evenly, each taking 12 cents. Abby says that since she supplied fifteen cookies and Betty only nine, she should take 15 cents and Betty 9 cents.

What is the fair division of the 24 cents between Abby and Betty?

Solution

The key is to determine the worth of one cookie. Each girl eats eight cookies. Since Carly pays 24 cents for eight cookies, each cookie is worth 3 cents. Thus Abby, who starts with fifteen cookies and sells seven to Carly, should receive 21 cents, and Betty, who starts with nine cookies and sells one to Carly, should receive 3 cents.

Bonus: Cookie Jar Division

Abby, Betty, and Carly have 21 cookie jars (all the same size). Seven are full, seven are half-full, and seven are empty. The girls wish to divide the cookie jars among themselves so that each girl gets the same number of jars and the same amount of cookies. How can they do this without opening the jars?

There are two different ways, as shown below.

Abby	F	F	F	H	E	E	E	Abby	F	F	H	H	H E E
Betty	F	F	F	H	E	E	E	Betty	F	F	H	H	H E E
Carly	F	H	H	H	H	H	E	Carly	F	F	F	H	E E E

We have designated the full cookie jars by F, the half-full ones by H, and the empty ones by E. In both solutions, each girl receives seven jars and $3\frac{1}{2}$ jars' worth of cookies. We have not counted as different the same solution with the girls' names permuted.

As we will see later (in "Integer Triangles" on p. 165), each solution corresponds to a triangle with integer sides and perimeter 7, as shown below. In a given triangle, the side lengths equal the number of full cookie jars for each girl in the corresponding cookie jar solution.

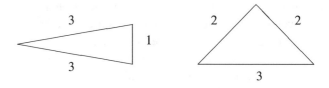

A Mere Fraction

(a) Find an integral fraction between 1/4 and 1/3 such that the denominator is a positive integer less than 10.

(b) Find an integral fraction between 7/10 and 5/7 such that the denominator is a positive integer less than 20.

Solution

(a) Since $3 < 3\frac{1}{2} < 4$, we have, upon taking reciprocals, the inequalities

$$\frac{1}{4} < \frac{2}{7} < \frac{1}{3}.$$

We double-check, by cross-multiplication:

$$\frac{1}{4} < \frac{2}{7} \quad \text{since} \quad 1 \times 7 < 4 \times 2$$

and

$$\frac{2}{7} < \frac{1}{3} \quad \text{since} \quad 2 \times 3 < 7 \times 1.$$

One can check by exhaustion that 2/7 is the only solution.

(b) Notice that the fraction found in (a) can be obtained by adding the numerators and denominators of $1/4$ and $1/3$:
$$\frac{1+1}{4+3} = \frac{2}{7}.$$
Would the same trick work for $7/10$ and $5/7$? Let's try the plausible answer
$$\frac{7+5}{10+7} = \frac{12}{17}.$$
We verify the inequalities
$$\frac{7}{10} < \frac{12}{17} < \frac{5}{7}$$
by cross-multiplication:
$$\frac{7}{10} < \frac{12}{17} \quad \text{since} \quad 7 \times 17 < 10 \times 12$$
and
$$\frac{12}{17} < \frac{5}{7} \quad \text{since} \quad 12 \times 7 < 17 \times 5.$$
One can check by exhaustion that $12/17$ is the only solution.

Bonus: Mediant Fractions

The answers to (a) and (b) are called *mediant fractions*.[1] The mediant fraction of a/b and c/d is $(a+c)/(b+d)$. Thus, the mediant fraction of $1/4$ and $1/3$ is $2/7$, and the mediant fraction of $7/10$ and $5/7$ is $12/17$. If $a/b < c/d$ (with b and d positive), then
$$\frac{a}{b} < \frac{a+c}{b+d} < \frac{c}{d}.$$
I encourage the reader to prove these inequalities using cross-multiplication.

Here is an aha! proof of the mediant fraction inequalities. Assuming that a, b, c and d are all positive, we will interpret the fractions as concentrations of salt in water. Suppose that we have two solutions of salt water, the first with a teaspoons of salt in b gallons of water, and the second with c teaspoons of salt in d gallons of water. The concentration of salt in the first solution is a/b teaspoons/gallon, while the concentration of salt in the second solution is c/d teaspoons/gallon. Suppose that the first solution is less salty than the second, i.e., $a/b < c/d$. Now, if we combine the two solutions, we obtain a solution with $a+c$ teaspoons of salt in $b+d$ gallons of water, so that the salinity of the new solution is $(a+c)/(b+d)$ teaspoons/gallon. Certainly, the new solution is saltier than the first solution and less salty than the second. That is to say,
$$\frac{a}{b} < \frac{a+c}{b+d} < \frac{c}{d}.$$
Our proof can truly be called a saline solution!

---

[1] Mediant fractions arise in the study of Farey sequences and in the "solution" of Simpson's Paradox.

A Long Sum

What is the sum of the first 100 integers,

$$1 + 2 + 3 + \cdots + 100?$$

Of course, we could laboriously add up the numbers. But we seek instead an aha! solution, a simple calculation that immediately gives the answer *and insight into the problem*.

Solution

Observe that the numbers may be paired as follows:

$$1 \text{ and } 100,$$
$$2 \text{ and } 99,$$
$$3 \text{ and } 98,$$
$$\ldots,$$
$$50 \text{ and } 51.$$

We have 50 pairs, each pair adding up to 101, so our sum is $50 \times 101 = 5050$.

This solution works in general for the sum

$$1 + 2 + 3 + \cdots + n,$$

where n is an even number. We pair the numbers as before:

$$1 \text{ and } n,$$
$$2 \text{ and } n - 1,$$
$$3 \text{ and } n - 2,$$
$$\ldots,$$
$$n/2 \text{ and } n/2 + 1.$$

We have $n/2$ pairs, each adding up to $n + 1$, so our sum is $n/2 \times (n + 1) = n(n + 1)/2$.

What if n is odd? We can no longer pair all the numbers (as the middle term has no mate). However, throwing in a 0 doesn't change the total:

$$0 + 1 + 2 + 3 + \cdots + n.$$

Now we have an even number of terms, and they may be paired as

$$0 \text{ and } n,$$
$$1 \text{ and } n - 1,$$
$$2 \text{ and } n - 2,$$
$$\ldots,$$
$$(n - 1)/2 \text{ and } (n - 1)/2 + 1.$$

We have $(n + 1)/2$ pairs, each adding up to n, so our sum is $n(n + 1)/2$ (again).

A "duplication method" works for both n even and n odd. Let S be the sum, and introduce a duplicate of S, written backwards:

S	$=$	1	$+$	2	$+$	3	$+$	$\cdots$	$+$	$n.$
S	$=$	n	$+$	$(n-1)$	$+$	$(n-2)$	$+$	$\cdots$	$+$	$1.$

1.1 Arithmetic

Add the two expressions for S, summing the first terms, then the second terms, and so on:

$$2S = (n+1) + (n+1) + (n+1) + \cdots + (n+1).$$

(The term $n+1$ occurs n times.) This simplifies to

$$2S = n(n+1),$$

or

$$S = \frac{n(n+1)}{2}.$$

Bonus: Sum of an Arithmetic Progression

Carl Friedrich Gauss (1777–1855), one of the greatest mathematicians of all time, is said to have solved our problem for $n = 100$ when he was a 10-year-old school pupil. However, according to E. T. Bell [2], the problem that Gauss solved was actually more difficult:

> The problem was of the following sort, $81297 + 81495 + 81693 + \ldots + 100899$, where the step from one number to the next is the same all along (here 198), and a given number of terms (here 100) are to be added.

The problem mentioned by Bell is the sum of an *arithmetic progression*. We can evaluate the sum using our formula for the sum of the first n integers. The calculation is

$$81297 + 81495 + \cdots + 100899 = 81297 \times 100 + 198 \times (1 + \cdots + 99)$$

$$= 8129700 + 198 \times \frac{99 \times 100}{2}$$

$$= 9109800.$$

Sums of Consecutive Integers

Behold the identities

$$1 + 2 = 3$$

$$4 + 5 + 6 = 7 + 8$$

$$9 + 10 + 11 + 12 = 13 + 14 + 15$$

$$16 + 17 + 18 + 19 + 20 = 21 + 22 + 23 + 24$$

$$25 + 26 + 27 + 28 + 29 + 30 = 31 + 32 + 33 + 34 + 35$$

$$36 + 37 + 38 + 39 + 40 + 41 + 42 = 43 + 44 + 45 + 46 + 47 + 48.$$

What is the pattern and why does it work?[2]

---

[2] Roger B. Nelsen gives a "proof without words" for this problem in the February 1990 issue of *Mathematics Magazine*.

Solution

Consider the third identity:

$$9 + 10 + 11 + 12 = 13 + 14 + 15.$$

If we add 4 to each of the first three numbers on the left (9, 10, and 11), then we obtain the three numbers on the right (13, 14, and 15). We have added $4 \cdot 3 = 12$ on the left, which is the fourth number on the left.

The nth identity (for $n \geq 1$) is

$$n^2 + (n^2 + 1) + \cdots + (n^2 + n) = (n^2 + n + 1) + (n^2 + n + 2) + \cdots + (n^2 + n + n).$$

There are $n + 1$ terms on the left and n terms on the right. If we add $n + 1$ to each term on the left except the last term, then we obtain all the terms on the right. We have added $(n + 1)n = n^2 + n$ on the left, and this is the last term on the left.

Bonus: Finding a Polynomial

It's easy to find the sum given by the nth identity. The average of the $n + 1$ terms on the left is $(n^2 + (n^2 + n))/2$, so the sum is

$$(n + 1)(n^2 + (n^2 + 1))/2 = n(n + 1)(2n + 1)/2.$$

Let's suppose that we didn't know this formula, but only the sums:

$$3, \; 15, \; 42, \; 90, \; 165, \; 273, \; \ldots.$$

How can we find the polynomial $p(n)$, whose values for $n = 1, 2, 3, \ldots$ are these numbers? The method (from finite difference calculus) is to make a sequence of differences of consecutive values of our starting sequence:

$$12, \; 27, \; 48, \; 75, \; 108, \; \ldots.$$

We repeat this process, creating a sequence of sequences:

$$
\begin{array}{cccccc}
3, & 15, & 42, & 90, & 165, & 273, & \ldots \\
12, & 27, & 48, & 75, & 108, & & \ldots \\
15, & 21, & 27, & 33, & & & \ldots \\
6, & 6, & 6, & & & & \ldots
\end{array}
$$

Having obtained a constant sequence, we stop. Now, the polynomial $p(n)$ is defined by multiplying the first column of our array by successive binomial coefficients and adding:

$$p(n) = 3\binom{n}{0} + 12\binom{n}{1} + 15\binom{n}{2} + 6\binom{n}{3} = \frac{(n+1)(n+2)(2n+3)}{2}.$$

This polynomial gives the values of our sequence starting at $p(0)$. Since we want to start at $p(1)$, we simply replace n by $n - 1$ to obtain the polynomial $p(n) = n(n+1)(2n+1)/2$.

Pluses and Minuses

Evaluate
$$100^2 - 99^2 + 98^2 - 97^2 + 96^2 - 95^2 + \cdots + 2^2 - 1^2.$$

Solution

Using the formula for the difference of two squares, $x^2 - y^2 = (x + y)(x - y)$, the expression can be written as

$$(100 + 99)(100 - 99) + (98 + 97)(98 - 97) + (96 + 95)(96 - 95) + \cdots + (2 + 1)(2 - 1).$$

Since every other term in parentheses is equal to 1, this expression simplifies to

$$100 + 99 + 98 + 97 + 96 + 95 + \cdots + 2 + 1.$$

In "A Long Sum," we found this sum to be 5050.

Bonus: Curious Identities.[3]

Can you explain the pattern for the identities

$$3^2 + 4^2 = 5^2$$
$$10^2 + 11^2 + 12^2 = 13^2 + 14^2$$
$$21^2 + 22^2 + 23^2 + 24^2 = 25^2 + 26^2 + 27^2$$
$$36^2 + 37^2 + 38^2 + 39^2 + 40^2 = 41^2 + 42^2 + 43^2 + 44^2 \,?$$

Let's prove the last one in a way that indicates what's going on in general. Transposing all terms on the left except 36^2 to the right side of the equation yields

$$36^2 = (41^2 - 40^2) + (42^2 - 39^2) + (43^2 - 38^2) + (44^2 - 37^2).$$

Notice that we paired the largest and smallest terms, the second-largest and second-smallest, etc. Now, using our difference of squares formula, we have

$$\begin{aligned} 36^2 &= (41 + 40)(41 - 40) + (42 + 39)(42 - 39) \\ &\quad + (43 + 38)(43 - 38) + (44 + 37)(44 - 37) \\ &= 81 \cdot 1 + 81 \cdot 3 + 81 \cdot 5 + 81 \cdot 7 \\ &= 81(1 + 3 + 5 + 7) \\ &= 81 \cdot 16. \end{aligned}$$

And certainly, $36^2 = 9^2 \cdot 4^2 = 81 \cdot 16$. Our computation tells the story in general. For $n \geq 1$, we claim that

$$[n(2n + 1)]^2 + \cdots + [2n(n + 1)]^2 = (2n^2 + 2n + 1)^2 + \cdots + (2n^2 + 3n)^2.$$

---

[3] Michael Boardman gives a "proof without words" for this problem in the February 2000 issue of *Mathematics Magazine*.

Transposing terms as we did in our test case, we obtain

$$[n(2n+1)]^2 = [(2n^2+2n+1)^2 - (2n(n+1))^2] + \cdots$$
$$+ [(2n^2+3n)^2 - (n(2n+1)+1)^2]$$
$$= (4n^2+4n+1)(1+3+5+7+\cdots+2n-1).$$

We need a formula for the sum of the first n odd numbers,

$$1+3+5+7+\cdots+2n-1.$$

Since this is the sum of an arithmetic progression, we could use the method in "A Long Sum." However, by the diagram below, we can see that the sum is equal to n^2.

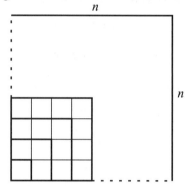

Using our formula for the sum of the first n odd numbers, our identity becomes

$$[n(2n+1)]^2 = (2n+1)^2 n^2,$$

which is obviously true. Our identity is verified!

Which is Greater?

Which number is greater,

$$\sqrt{6}+\sqrt{10} \quad \text{or} \quad \sqrt{5}+\sqrt{12}\,?$$

We could compute square roots, but we seek instead an aha! solution that gives the answer immediately.

Solution

The winning idea is to compare the squares of the two numbers (thus eliminating some of the square roots). The larger number has the larger square.

The squares of the given numbers are

$$\left(\sqrt{6}+\sqrt{10}\right)^2 = 6 + 2\sqrt{60} + 10 = 16 + 2\sqrt{60}$$

and

$$\left(\sqrt{5}+\sqrt{12}\right)^2 = 5 + 2\sqrt{60} + 12 = 17 + 2\sqrt{60}.$$

The second square is larger. Therefore, $\sqrt{5}+\sqrt{12}$ is greater than $\sqrt{6}+\sqrt{10}$.

Bonus: A Diophantine Equation

The numbers in this problem—let's call the smaller one x and the larger one y—satisfy the equation
$$x^2 + 1 = y^2.$$

This is an example of a *Diophantine equation*, after the Greek mathematician Diophantus (c. 200–c. 284). Diophantus called for rational number solutions to such equations. In our problem, x and y are not rational numbers (as their squares are irrational numbers). Let's find all rational solutions to the equation.

The graph of the equation is a hyperbola, as shown in the picture.

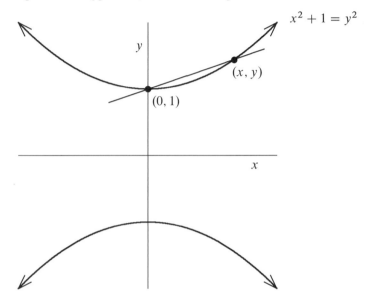

The point $(0, 1)$ is on the hyperbola, so we have one rational solution. If (x, y) is another rational solution (where $x \neq 0$), then the slope of the line through $(0, 1)$ and (x, y) is also rational. Let's call this slope m. Thus
$$m = \frac{y - 1}{x - 0}.$$

Solving for y, we obtain $y = mx + 1$. Substituting this expression for y into the equation of the hyperbola yields
$$x^2 + 1 = (mx + 1)^2$$
$$= m^2 x^2 + 2mx + 1.$$

Solving for x we have
$$x = \frac{2m}{1 - m^2},$$
and hence
$$y = \frac{1 + m^2}{1 - m^2}.$$

These equations, where m is a rational number not equal to ± 1, are a parameterization of all rational solutions to the equation $x^2 + 1 = y^2$, except for the solution $(0, -1)$. For example, if $m = 4/11$, then $(x, y) = (88/105, 137/105)$. The reason that $(0, -1)$ isn't included is because it would determine a line with undefined slope. The value $m = 0$ corresponds to a tangent line to the hyperbola at the point $(0, 1)$. Can you see which values of m correspond to the upper and lower branches of the hyperbola?

Cut Down to Size

Find two whole numbers greater than 1 whose product is 999,991.

Solution

Observe that

$$999{,}991 = 1{,}000{,}000 - 9$$
$$= 1000^2 - 3^2.$$

The algebra rule for factoring the difference of two squares comes in handy:

$$x^2 - y^2 = (x - y)(x + y).$$

Applying the rule with $x = 1000$ and $y = 3$, we obtain

$$999{,}991 = 1000^2 - 3^2$$
$$= (1000 - 3)(1000 + 3)$$
$$= 997 \times 1003.$$

Bonus: Finding the Prime Factorization

We have found two numbers, 997 and 1003, whose product is 999,991. Can these factors be broken down further (i.e., do they have proper divisors?), or are they prime numbers (see Toolkit)? We can divide 997 and 1003 by various other numbers, such as 2, 3, etc., to see if the quotients are integers, but how many trials would we have to make?

When testing for proper divisors of n, we need only divide n by prime numbers, because if n has a proper divisor then that divisor has a prime factor. As we test possible divisors, 2, 3, 5, etc., the quotients, $n/2$, $n/3$, $n/5$, etc., become smaller. The "break-even point" occurs at $\sqrt{n}$, since $n/\sqrt{n} = \sqrt{n}$. Hence we only need to search for prime divisors up to $\sqrt{n}$. If there are no such divisors, then n is a prime number.

As $31 < \sqrt{997} < 32$, the only prime numbers we need to check as possible divisors of 997 are 2, 3, 5, 7, 11, 13, 17, 19, 23, 29, and 31. Upon division, we find that none of these primes divides 997 evenly, so 997 is a prime number. To factor 1003, we work with the same set of primes, since $31 < \sqrt{1003} < 32$. Checking these, we hit pay-dirt with 17 and find that $1003 = 17 \times 59$, the product of two primes. To verify that 59 is a prime, you need only check that 59 isn't divisible by 2, 3, 5, or 7, since $7 < \sqrt{59} < 8$.

Therefore, the prime factorization of 999,991 is $17 \times 59 \times 997$.

Ordered Digits

How many positive integers have the property that their digits increase as read left-to-right? Examples: 19, 357, and 2589.

Solution

Each nonempty subset of the set of integers $\{1, 2, 3, 4, 5, 6, 7, 8, 9\}$ yields such a number. For example, the subset $\{2, 5, 8, 9\}$ yields the integer 2589. An n-element set has 2^n subsets (including the empty set). In our problem, $n = 9$ and we exclude the empty set, so there are $2^9 - 1 = 511$ such numbers.

Bonus: Ordered Digits in a Square

What square numbers have the property that their digits are in nondecreasing order read left-to-right? Examples: $12^2 = 144$, $13^2 = 169$, and $83^2 = 6889$.

Donald Knuth attacked this problem in one of his 1985 "Aha sessions," which were classes where he and his students worked on challenging problems. You can find videos of the sessions on the Stanford Center for Professional Development web pages (http://scpd.stanford.edu/knuth/).

Let's show the existence of an infinite collection of perfect squares whose digits are in order. This was found by Anil Gangolli.

We will demonstrate that

$$(\underbrace{6\ldots 6\,7}_{n})^2 = \underbrace{4\ldots 4}_{n+1}\underbrace{8\ldots 8}_{n}9, \quad n \geq 1.$$

For example, the case $n = 1$ is the statement that $67^2 = 4489$.

Since

$$\underbrace{6\ldots 6\,7}_{n} = \frac{2}{3}10^{n+1} + \frac{1}{3},$$

we have

$$(\underbrace{6\ldots 6\,7}_{n})^2 = \frac{4}{9}10^{2n+2} + \frac{4}{9}10^{n+1} + \frac{1}{9}$$

$$= \underbrace{4\ldots 4}_{2n+2} + \frac{4}{9} + \underbrace{4\ldots 4}_{n+1} + \frac{4}{9} + \frac{1}{9}$$

$$= \underbrace{4\ldots 4}_{n+1}\underbrace{8\ldots 8}_{n}9.$$

A similar infinite family is:

$$(\underbrace{3\ldots 3\,4}_{n})^2 = \underbrace{1\ldots 1}_{n+1}\underbrace{5\ldots 5}_{n}6, \quad n \geq 1.$$

There are many other infinite families with the desired property.

What's the Next Term?

Give the next term in each of the following sequences:

(a)
$$1, 4, 9, 16, 25, 36, 49, 64, 81, \ldots$$

(b)
$$0, 1, 1, 2, 3, 5, 8, 13, 21, 34, 55, 89, \ldots$$

(c)
$$1, 1, 1, 1, 2, 1, 1, 3, 3, 1, 1, 4, 6, 4, 1, 1, 5, \ldots$$

(d)
$$0, 1, 2, 2, 3, 3, 4, 4, 4, 4, 5, 5, 6, 6, 6, 6, 7, 7, \ldots$$

(e)
$$1, 2, 4, 6, 16, 12, 64, 24, 36, 48, 1024, 60, \ldots.$$

Solution

For each sequence, try to relate the numbers to a pattern that you have seen before.

(a) The terms are the square numbers, n^2. So the next term is $10^2 = 100$.

(b) The terms are those of the famous Fibonacci sequence, $\{f_n\}$, defined by $f_0 = 0$, $f_1 = 1$, and $f_n = f_{n-1} + f_{n-2}$, for $n \geq 2$. So the next term is 144.

(c) The terms are the entries in Pascal's triangle (see Toolkit), reading across and down. So the next term is 10.

(d) The terms are the values of $\pi(n)$, the number of primes less than or equal to n. So the next term is $\pi(19) = 8$.

(e) The terms are the smallest numbers with n positive divisors. So the next term is the smallest number with 13 positive divisors. This number is $2^{12} = 4096$.

Bonus: The On-Line Encyclopedia of Integer Sequences

A good resource for tracking down an integer sequence is "The On-Line Encyclopedia of Integer Sequences," presided over by Neil J. A. Sloane (the on-line address is www.research.att.com/~njas/sequences/). Simply type the first terms of a sequence that you are interested in, press enter, and the search engine will show you sequences that match. For example, if we enter the terms from (e) above,

$$1, 2, 4, 6, 16, 12, 64, 24, 36, 48, 1024, 60,$$

then the search comes up with sequence A005179 (the smallest number with n divisors).

1.2 Algebra

How Does She Know?

A mother asks her daughter to choose a number between 1 to 10 but not reveal the number. She then gives her daughter the following directions: add 7 to the number; double the result; subtract 8; divide by 2; subtract the original number. She tells her daughter that the final result is 3. How does the mother know this?

Solution

Suppose that the daughter chooses the number x (which can be any number). Then the steps can be written algebraically as:

choose a number	x
add 7	$x + 7$
double the result	$2(x + 7) = 2x + 14$
subtract 8	$2x + 14 - 8 = 2x + 6$
divide by 2	$(2x + 6)/2 = x + 3$
subtract the original number	$x + 3 - x = 3.$

Thus, the initial number x has disappeared from the computation and the final result is 3, regardless of x.

As an algebraic identity, we are simply saying

$$\frac{2(x + 7) - 8}{2} - x = 3.$$

Bonus: Clock Magic

Ask a friend to think of any number on a clock (a clock with twelve numbers) without revealing the number. Tell her that you are going to point to different numbers on the clock while she silently counts up to 20, starting with her secret number and adding 1 every time you point to a number. When she reaches 20, she should say "stop." You can arrange it so that when she says "stop," the number you are pointing to is the number she chose. How do you do this?

Keeping a silent count, you begin by pointing to seven numbers on the clock at random. (Your friend cannot have said "stop" already because $20 - 7 = 13$, and 13 is not a number on the clock.) Then you point to the 12, then the 11, then the 10, and so on, going backwards around the clock, until your friend says "stop." After the first seven numbers, your count plus the number that you are currently pointing to always add up to 20 (e.g., $8 + 12$, $9 + 11$, $10 + 10$). So when your friend says "stop," the number that you are pointing to is equal to 20 minus your count, and this is the number that your friend started her count with. For example, if your friend is thinking of 7:00, then you will point to thirteen numbers altogether ($20 - 7 = 13$), and the last number that you point to will be $20 - 13 = 7$.

In algebraic terms, let

$$f = \text{friend's number},$$

$$y = \text{your count},$$

$$p = \text{number you are pointing to (after the first seven)}.$$

Note that y and p are not constants; they change with each step of the count. We always have

$$p = 20 - y.$$

Hence, when your friend says "stop,"

$$f = 20 - y = p.$$

How Cold Was It?

It was so cold that Fahrenheit equaled Centigrade. How cold was it?

Solution

We seek a number of degrees that represents the same temperature in Centigrade and Fahrenheit. We can solve this problem if we know the freezing temperature and the boiling temperature of water in both Centigrade (C) and Fahrenheit (F). As shown in the table below, water freezes at 0° C and 32° F, and it boils at 100° C and 212° F.

	C	F
water freezes	0°	32°
water boils	100°	212°

We observe from the table that a change in temperature of 100° C is equivalent to a change in temperature of 180° F. So a change in temperature of 10° C is equivalent to a change in temperature of 18° F. Hence,

$$-10° \text{ C} = 14° \text{ F}$$

$$-20° \text{ C} = -4° \text{ F}$$

$$-30° \text{ C} = -22° \text{ F}$$

$$-40° \text{ C} = -40° \text{ F}.$$

Therefore, $-40°$ represents the same temperature in Centigrade and Fahrenheit. The answer is unique since as the temperature moves away from $-40°$, the Fahrenheit reading changes by more than the Centigrade reading.

Bonus: Centigrade to Fahrenheit

Let's find the conversion formula from c degrees Centigrade to f degrees Fahrenheit.[4] The proportional change of temperature found in the Solution may be written

$$\frac{(c-0)°}{100°} = \frac{(f-32)°}{180°}.$$

Upon simplification, we obtain the linear relation

$$(9/5)c° + 32° = f°.$$

For example, the average normal human body temperature is about 37° C, so this is

$$(9/5)37° + 32° = 99° \text{ F (to two significant digits).}$$

Man vs. Train

A man is crossing a train trestle on foot. When he is $4/7$ of the way across he sees a train coming toward him head-on. He realizes that he has just enough time to run toward the train and get off the trestle or to run away from the train and get off the trestle. If the man can run 20 kilometers per hour, how fast is the train going?

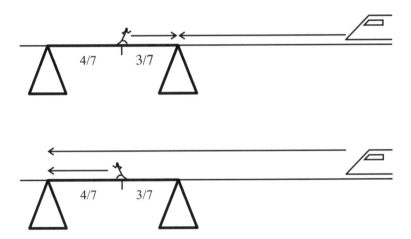

Solution

It's easy to get bogged down with formulas involving distances, rates, etc. A simple solution obviates the need for complicated calculations.

The solution hinges on the fact that the man and train can be together at two points. Let's distinguish between the near end and the far end of the trestle. The man can meet the train at the near end after running $3/7$ of the length of the trestle. If instead he runs this same distance toward the far end, then the train will be at the near end at the same time as he is $1/7$ of the length of the trestle from the far end. Since he can meet the train at the far end, the train goes 7 times faster than he does, or 140 kilometers per hour.

---

[4]The Centigrade temperature scale was invented in 1742 by the astronomer Anders Celsius (1701–1744). The Fahrenheit scale was invented around 1724 by the physicist Daniel Fahrenheit (1686–1736).

Bonus: Sum of a Geometric Series

Speaking of trains and simple solutions, the story goes that John von Neumann[5] was asked a problem of the following type: Two trains are headed toward each other on the same track, each traveling at 60 miles per hour. When they are 2 miles apart, a fly leaves the front of one train and travels at 90 miles per hour to the front of the other train. (Flies don't really fly this fast.) Then it travels back to the first train, and so on, back and forth until the two trains crash. How far does the fly travel?

There is a quick way and a methodical way to solve this problem. The quick way is to realize that the trains crash in 1 minute (since they start 2 miles apart and are traveling at the rate of 1 mile per minute). Since the fly travels at the rate of 1.5 miles per minute, it travels 1.5 miles in this time.

The methodical way to solve the problem is to sum an infinite geometric series. The fly completes the first step of its journey (traveling from one train to the other) in $4/5$ of a minute, since in this time the oncoming train travels $4/5$ of a mile and the fly travels $6/5$ of a mile. Therefore, after $4/5$ of a minute, we have a similar version of the original problem, but with the trains $2 - 2 \times 4/5 = 2/5$ miles, or $1/5$ of the original distance, apart. This pattern continues, forming an infinite geometric series of distances, the sum of which is the total distance the fly travels:

$$\frac{6}{5}\left(1 + \frac{1}{5} + \left(\frac{1}{5}\right)^2 + \left(\frac{1}{5}\right)^3 + \cdots\right) \text{ miles.}$$

Let's find the sum of the infinite geometric series

$$S = 1 + r + r^2 + r^3 + \cdots,$$

where r is a real number with $-1 < r < 1$. (We need these bounds on r so that the series converges; more on this in a moment.) Multiplying by r, we obtain

$$Sr = r + r^2 + r^3 + r^4 + \cdots.$$

Subtracting the second equation from the first yields

$$S - Sr = 1,$$

or

$$S(1 - r) = 1,$$

and so

$$S = \frac{1}{1 - r}.$$

Letting $r = 1/5$, we complete our calculation of the distance traveled by the fly:

$$\frac{6}{5} \times \frac{1}{1 - 1/5} = \frac{3}{2} \text{ miles}$$

(the same answer as before).

---
[5] John von Neumann (1903–1957) made contributions in several areas of mathematics, including computer science and game theory.

1.2 Algebra

Von Neumann gave the correct answer instantly, so the questioner said that he surely must have calculated the time the fly travels (our first method). Von Neumann replied that, on the contrary, he had summed the series.

By the way, we can also obtain the sum of a finite geometric series,

$$S = 1 + r + r^2 + r^3 + \cdots + r^n,$$

where r is any real number not equal to 1. As before, multiply S by r, take the difference of the two equations, and solve for S. This yields

$$S = \frac{1 - r^{n+1}}{1 - r}.$$

The convergence of the infinite geometric series is based on what happens to the finite geometric series as n increases. If $-1 < r < 1$, then the term r^{n+1} gets closer to 0 as n gets larger. Hence, as n tends to infinity, the sum of the finite geometric series tends to $1/(1 - r)$, which is our formula for the sum of the infinite geometric series. In this sense, we say that the infinite series converges (and the sum is given by our formula).

The sum with $r = 1/2$, i.e.,

$$1 + \frac{1}{2} + \frac{1}{4} + \frac{1}{8} + \cdots = \frac{1}{1 - 1/2} = 2$$

is particularly memorable. We see in the picture below that a rectangle of dimensions 1×2 is completely filled by smaller rectangles whose areas are the geometrically decreasing terms of the series.

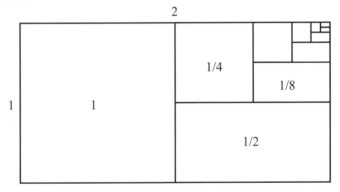

Uphill, Downhill

A bicyclist goes up a hill at 30 km/hr and down the same hill at 90 km/hr. What is the cyclist's average speed for the trip?

Solution

At first, you might think that the answer is the simple average of 30 km/hr and 90 km/hr, i.e., 60 km/hr. But this isn't correct since the cyclist spends less time at the faster rate.

A quick way to find the average speed is to assume that the answer is independent of the length of the hill. If that's true, then we can set the length of the hill to a convenient value, say, 90 km. Then the trip takes 3 hours up the hill and 1 hour down. So the average speed is 180 km /4 hr = 45 km/hr.

To double-check this, suppose that the length of the hill is d km. Then the time going up the hill is $d/30$ hr and the time going down the hill is $d/90$ hr. Hence, the average speed is

$$\frac{2d}{\frac{d}{30} + \frac{d}{90}} = \frac{2}{\frac{1}{30} + \frac{1}{90}} = 45 \text{ km/hr}.$$

The average speed is called the *harmonic mean* of the two rates. As we have seen, the harmonic mean is less than the arithmetic mean of the two rates (see the Bonus).

Bonus: Power Means

For any n positive real numbers $x_1, x_2, \ldots, x_n$ and any real number p, we define the *p-power mean* by the formula

$$M_p = \left(\frac{x_1^p + x_2^p + \cdots + x_n^p}{n} \right)^{1/p},$$

for $p \neq 0$, and

$$M_0 = (x_1 x_2 \ldots x_n)^{1/n}.$$

For $p = 1, 0,$ and -1, these means are the arithmetic mean (AM), geometric mean (GM), and harmonic mean (HM), respectively.

For $x_1, x_2, \ldots, x_n$ fixed and not all equal, M_p is an increasing function of p. There is equality in the means if and only if all the x_i are equal. In particular, we have the AM–GM–HM inequalities: HM $\leq$ GM $\leq$ AM, with equality if and only if all the x_i are equal.

How Many Solutions?

How many solutions has the equation

$$x + 2y + 4z = 100,$$

where x, y, and z are nonnegative integers?

Solution

There are 26 choices for z, namely, all integers from 0 to 25. Among these choices, the average value of $4z$ is 50. So, on average, $x + 2y = 50$. In this equation, there are 26 choices for y, namely, all integers from 0 to 25. The value of x is determined by the value of y. Hence, altogether there are $26^2 = 676$ solutions to the original equation.

Bonus: Distributions, Partitions, and Schur's Estimate

By changing the coefficients of x, y, and z, we obtain many variations of our problem, all more difficult than the one we tackled.

If all the coefficients are 1, then we have the equation

$$x + y + z = 100,$$

which is called a *distribution*. The number of solutions in nonnegative integers is $\binom{102}{2} = 5151$ (see "Bonus: The Correct Number of Orders").

If the coefficients are 1, 2, and 3, we obtain the equation

$$x + 2y + 3z = 100,$$

which is a *partition* of 100 into three or fewer parts. The number of partitions of n into one part is 1. The number of partitions of n into two parts is $\lfloor n/2 \rfloor$. The number of partitions of n into three parts is $\{n^2/12\}$, where $\{\}$ denotes the nearest-integer function. Hence, the number of partitions of 100 into three or fewer parts is $1 + \lfloor 100/2 \rfloor + \{100^2/12\} = 1 + 50 + 833 = 884$.

Let a, b, and c be the coefficients of x, y, and z, respectively. If a, b, and c are relatively prime integers (they have no factors in common), then the number of nonnegative integer solutions to the equation

$$ax + by + cz = n$$

is approximated by Issai Schur's[6] asymptotic estimate

$$\frac{n^2}{2abc}.$$

1.3 Geometry

A Quadrilateral from a Quadrilateral

Let $ABCD$ be a quadrilateral. Let E, F, G, and H be the midpoints of sides AB, BC, CD, and DA, respectively. Prove that the area of the quadrilateral $EFGH$ is half the area of $ABCD$.

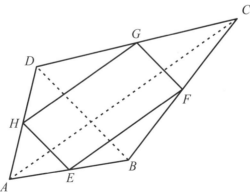

---

[6] Issai Schur (1875–1941) was a mathematician who worked primarily in algebra.

Solution

Assume that the quadrilateral is convex, as in the picture (the concave case is similar). Add the two construction lines AC and BD, as shown by dotted lines. By a simple theorem of geometry (see Bonus), the area of triangle AEH is one-fourth the area of triangle ABD. Likewise, the area of triangle CFG is one-fourth the area of triangle CBD. Hence, the area of triangles AEH and CFG together is one-fourth the area of the quadrilateral $ABCD$. Similarly, the area of triangles DGH and BEF together is one-fourth the area of $ABCD$. It follows that the area of the four triangles AEH, CFG, DGH, and BEF together is half the area of $ABCD$. Therefore, the area of the complement of these triangles, which is the quadrilateral $EFGH$, is half the area of $ABCD$.

Bonus: A Vector Proof

We give a quick vector proof that the area of triangle AEH is one-fourth the area of triangle ABD. Represent the points A, B, D, E, and H by vectors a, b, d, e, and h, respectively. Then $e = (a+b)/2$ and $h = (a+d)/2$. Hence

$$h - e = \frac{1}{2}(d - b).$$

Thus, the vector $h - e$ is parallel to the vector $d - b$ and half its length. Hence, the line segment EH is parallel to the line segment BD and half its length. It follows that triangle AEH is similar to triangle ABD, with a constant of proportionality of one-half. Therefore, the area of triangle AEH is one-fourth the area of triangle ABD.

Note. From the proof, we see that the quadrilateral $EFGH$ is a parallelogram. It is called the *Varignon parallelogram*, after Pierre Varignon.[7]

The Pythagorean Theorem

Prove: In any right triangle, the area of the square on the hypotenuse is equal to the sum of the areas of the squares on the other two sides. (See Toolkit.)

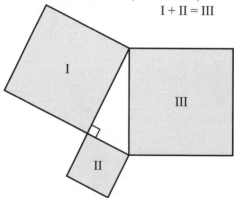

In the diagram, the areas satisfy the relation I + II = III.

This is half of the famous Pythagorean theorem. The other half is the converse statement: if I + II = III, then the angle opposite III is a right angle.

---
[7]Pierre Varignon (1654–1722) was a mathematical physicist and early advocate of calculus.

1.3 Geometry

Solution

Here is the most visually compelling proof that I know of. It requires very little explanation.

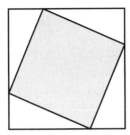 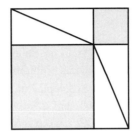

Both of the above figures show four copies of the given right triangle, arranged inside a square. The shaded area in the figure on the left is the square on the hypotenuse. Clearly, the shaded area doesn't change when the four triangles are rearranged as in the figure on the right. In this figure, the shaded area consists of the squares on the two legs of the triangle.

Bonus: A One-Triangle Proof

The above proof employs four copies of the given right triangle. Here is a proof that uses only the original right triangle.

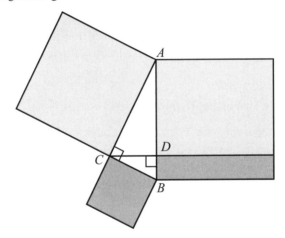

Let ABC be a right triangle with right angle C. Let D be the point on AB such that AB and CD are perpendicular. Extend the line CD so that it meets the square on the hypotenuse at a second point. (See the above figure.) By similar triangles, $|AB|/|BC| = |BC|/|BD|$, and hence $|AB| \cdot |BD| = |BC|^2$. It follows that the two dark gray areas are equal. Similarly, we can show that the two light gray areas are equal. Adding areas, we see that the area of the square on the hypotenuse is equal to the sum of the areas of the squares on the other two sides.

Nearly 100 proofs of the Pythagorean theorem are presented at "Cut-the-Knot" (http://www.cut-the-knot.org), a site "for teachers, parents and students who seek engaging mathematics."

Building Blocks

What are the proportions of a 30°–60°–90° triangle?

Solution

In the solution to "A Long Sum," we worked with a sum S written twice. Can we do something similar in this problem?

Let's work with two copies of the given triangle. Let the given triangle be ABC with $A = 30°$, $B = 60°$, and $C = 90°$. As shown in the diagram below, we put a copy of ABC, labeled $AB'C$, next to ABC.

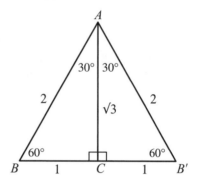

The larger triangle formed, ABB', is an equilateral triangle (it has three 60° angles). Hence, if $|BC| = 1$, then $|BB'| = 2$, and so also $|AB| = 2$. Now, by the Pythagorean theorem, $|AC| = \sqrt{3}$.

Bonus: Another Memorable Triangle

What are the proportions of a 36°–72°–72° triangle?

In our solution for the 30°–60°–90° triangle, we put two geometric pieces together. To solve the present problem, we do the reverse, dividing the triangle into two pieces. In the diagram below, ABC is a 36°–72°–72° triangle with $A = B = 72°$ and $C = 36°$. The triangle ABC is divided into two smaller triangles, another 36°–72°–72° triangle and a 36°–36°–108° triangle. Let $|AB| = 1$ and $|AC| = |BC| = \phi$. By similar triangles, ϕ satisfies the relation $\phi/1 = 1/(\phi - 1)$, so that $\phi^2 - \phi - 1 = 0$. From the quadratic formula (Toolkit), we find that $\phi = (\sqrt{5} + 1)/2$. Thus, ϕ is the famous "golden ratio" (more about ϕ in the problem "Irrational ϕ").

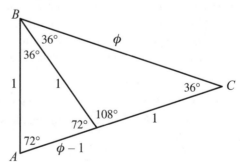

We see at once that

$$\cos 36° = \phi/2 = (\sqrt{5}+1)/4$$

and

$$\cos 72° = (\phi - 1)/2 = (\sqrt{5}-1)/4.$$

A Geometric Inequality

In the figure below, show that $a + b > 2c$.

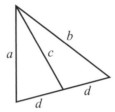

Solution

In the previous problem ("Building Blocks"), we saw a solution involving the construction of some additional lines. Let's try the same trick here.

Augment the figure to make a parallelogram with diagonals of lengths $2c$ and $2d$. Two adjacent sides of the parallelogram and the diagonal of length $2c$ make a triangle, so by the triangle inequality, $a + b > 2c$.

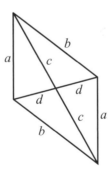

Bonus: A Squarable Lune

It is impossible to construct a square of the same area as a given circle, using straightedge and compass. However, the problem of "squaring a lune" is possible for certain lunes (a lune is formed by two intersecting circles of different radii).

Starting with the diagram below, let's construct a square with the same area as the shaded lune. This construction was discovered by Hippocrates of Chios (c. 470–c. 410 BCE).

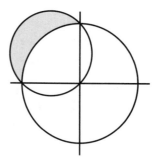

A number of auxiliary lines and circles are helpful. In the following diagram, by the Pythagorean theorem, the area of the semicircle on the hypotenuse of the inscribed right triangle is equal to the sum of the areas of the semicircles on the two sides.

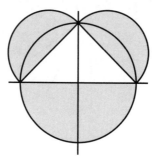

The shaded semicircle on the hypotenuse may just as well be moved to the upper half of the circle. This yields two heavily shaded overlapping areas, shown below, where the area is counted twice.

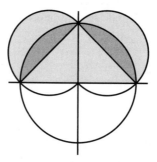

Subtracting overlapping areas, it follows that the area of the two lunes below is equal to the area of the inscribed right triangle.

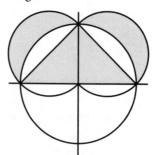

By symmetry, the area of one lune is equal to the area of one of the small right triangles.

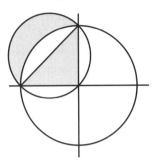

Now, it's easy to square the lune, as shown in the diagram below.

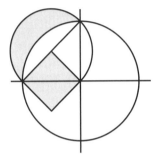

A Packing Problem

Find a way to pack (without overlap) 5 unit squares inside a square of side length less than 3.

Solution

The packing below of five unit squares in a square of side length $2 + 1/\sqrt{2} \approx 2.707$ was proved by Frits Göbel to be best-possible (the large square is as small as possible while still containing five unit squares).

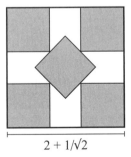

$2 + 1/\sqrt{2}$

Erich Friedman shows many packing configurations (squares in squares, circles in squares, etc.) at http://www.stetson.edu/~efriedma/packing.html.

Bonus: Covering With Unit Squares

Here is an unsolved covering problem: Prove that $n^2 + 1$ unit squares in a plane cannot cover a square of side length greater than n. (The perimeter and interior of the given square must be covered.) Below is an illustration of the problem where $n = 3$. The ten unit squares fail to cover a square of side length greater than 3.

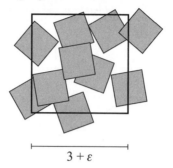

What's the Area?

Suppose that the large triangle below is equilateral with area 1. What is the area of the black region? (The black triangles form an infinitely nested pattern.)

Solution

Add dotted white lines as in the picture below.

1.3 Geometry

There are 16 subtriangles. Omitting the central one, we see that in the first stage we cover $12/15 = 4/5$ of the subtriangles. Since the pattern repeats, the black region has area $4/5$.

Bonus: Calculation by Geometric Series

At each stage we cover $3/4$ of the equilateral triangle, and the equilateral triangles decrease in area by a factor of 16. So by the formula for the sum of a geometric series (p. 16), the area is

$$\frac{3}{4}\left(1 + \frac{1}{16} + \frac{1}{16^2} + \cdots\right) = \frac{3}{4} \cdot \frac{1}{1 - 1/16} = \frac{4}{5}.$$

Volume of a Tetrahedron

Find the volume of a regular tetrahedron of edge length 1.

Solution

Place the vertices of the tetrahedron at four vertices of a cube with edge length $1/\sqrt{2}$. (These are opposite pairs of vertices on opposite faces of the cube.)

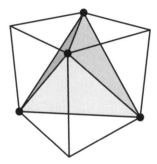

The volume of the tetrahedron is the volume of the cube minus the volumes of four identical isosceles right tetrahedra that are corners of the cube:

$$\left(\frac{1}{\sqrt{2}}\right)^3 - 4 \cdot \frac{1}{6}\left(\frac{1}{\sqrt{2}}\right)^3 = \frac{\sqrt{2}}{12}.$$

Using a similar argument, we can prove that the volume of a regular octahedron of edge length 1 is $\sqrt{2}/3$.

Bonus: Symmetry Groups

The way that the tetrahedron is inscribed in the cube above reveals a connection between the symmetry groups of the regular tetrahedron and the cube. If we pick up the cube and set it down again so that it occupies its original space, then the vertices, edges, and faces of the cube may have changed position. The symmetry group of the cube is the group of all such ways to reposition the cube. It's easy to find the order (number of elements) of the symmetry group of the cube. Since we can set the cube down on any of its six faces, and once we have done this we can rotate it in any of four ways, there are $6 \cdot 4 = 24$

symmetries in all. However, we still need to see which 24-element group this is. We know that the group of permutations on 4 objects has $4! = 24$ elements. This group is called the *symmetric group* of order 4, denoted S_4. In fact, the group of symmetries of a cube is isomorphic (equivalent) to S_4. To see this, we just need to find four parts of the cube that are permuted in all possible ways by the symmetries of the cube. The four diagonals of the cube have this property. Every symmetry of the cube permutes (interchanges) the diagonals, and conversely every possible permutation of the diagonals comes from a symmetry of the cube.

Similarly, we can show that the symmetry group of a regular octahedron is S_4.

Now that we know the symmetry group of the cube (S_4), we find that the symmetry group of the regular tetrahedron goes along for the ride. Every symmetry of the cube automatically gives a symmetry of the regular tetrahedron (with the tetrahedron inscribed in the cube), or it moves the vertices of the tetrahedron to the other four vertices of the cube. How many symmetries of the regular tetrahedron are there? Since we can put the tetrahedron down on any of its four faces and then rotate the tetrahedron in any of three ways, the regular tetrahedron has $4 \cdot 3 = 12$ symmetries. The symmetries of the tetrahedron comprise a subgroup of S_4 of order 12. It can be shown that this subgroup is the *alternating group* A_4, consisting of "even permutations" in S_4.

Irrational ϕ

Show that the golden ratio, $\phi = (\sqrt{5} + 1)/2$, is an irrational number.

Solution

The golden ratio ϕ is perhaps the irrational number with the simplest proof of irrationality. The definition of ϕ comes from the relation $\phi = 1/(\phi - 1)$ that we saw in the "Building Blocks" problem. A rectangle with dimensions $\phi \times 1$ has the property that if a square is removed, the remaining rectangle has the same proportions as the original.

We will prove by contradiction that ϕ is irrational. Suppose that ϕ is a rational number, say, n/m, where m and n are positive integers. Then we can construct such a rectangle with sides m and n (see the diagram).

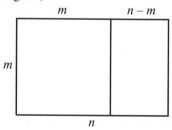

Now the relation $\phi = 1/(\phi - 1)$ implies that

$$\frac{n}{m} = \frac{1}{n/m - 1} = \frac{m}{n - m}.$$

Hence, ϕ is equal to a fraction, $m/(n - m)$, with smaller positive integral numerator and denominator. But this process could be repeated forever, which is clearly impossible, since

1.3 Geometry

we started with integers m and n. Therefore, our assumption that ϕ is rational is false; it is irrational.

Bonus: Irrationality of $\sqrt{2}$

There are several proofs that $\sqrt{2}$ is an irrational number. One well-known proof involves investigating the parity of the numerator and denominator of a supposed rational representation of $\sqrt{2}$. By contrast, here is a proof using an infinite regression argument.

Suppose that $\sqrt{2} = p/q$, where p and q are positive integers. Then $p^2 = 2q^2$, and so

$$\sqrt{2} = \frac{p}{q} = \frac{p(p-q)}{q(p-q)} = \frac{p^2 - pq}{q(p-q)} = \frac{2q^2 - pq}{q(p-q)} = \frac{q(2q-p)}{q(p-q)} = \frac{2q-p}{p-q}.$$

Since $q < p < 2q$, we see that $0 < 2q - p < p$ and $0 < p - q < p$. Hence, we have a pair of positive integers, $2q - p$ and $p - q$, which are smaller than p and q, respectively, and their quotient is $\sqrt{2}$. But this process could be repeated forever, producing smaller and smaller positive integers whose quotient is equal to $\sqrt{2}$. Clearly, this is impossible. Therefore $\sqrt{2}$ is irrational.

Tangent of a Sum

In trigonometry, the formula for the tangent of a sum of two angles is

$$\tan(A + B) = \frac{\tan A + \tan B}{1 - \tan A \tan B}.$$

Find a formula for $\tan(A + B + C + D + E)$ in terms of the tangents of the angles A, B, C, D, and E.

Solution

We start by finding a formula for the tangent of the sum of three angles. Using the formula given for the tangent of the sum of two angles, we have

$$\tan(A + B + C) = \frac{\tan(A+B) + \tan C}{1 - \tan(A+B)\tan C}$$

$$= \frac{\frac{\tan A + \tan B}{1 - \tan A \tan B} + \tan C}{1 - \left(\frac{\tan A + \tan B}{1 - \tan A \tan B}\right)\tan C}$$

$$= \frac{\tan A + \tan B + (1 - \tan A \tan B)\tan C}{1 - \tan A \tan B - (\tan A + \tan B)\tan C}$$

$$= \frac{\tan A + \tan B + \tan C - \tan A \tan B \tan C}{1 - \tan A \tan B - \tan A \tan C - \tan B \tan C}.$$

Notice that the numerator in this formula contains the sum of $\tan A$, $\tan B$, and $\tan C$, as well as the product of these terms, while the denominator contains all products of pairs of these terms. Accordingly, we introduce notation for the sum of all products of k tangent terms, where $1 \leq k \leq 3$. These expressions are based on the *elementary symmetric polynomials* (see Bonus). Define

$$s_1(A, B, C) = \tan A + \tan B + \tan C,$$
$$s_2(A, B, C) = \tan A \tan B + \tan A \tan C + \tan B \tan C,$$
$$s_3(A, B, C) = \tan A \tan B \tan C.$$

Thus, we can write our formula for the tangent of the sum of three angles as

$$\tan(A + B + C) = \frac{s_1(A, B, C) - s_3(A, B, C)}{1 - s_2(A, B, C)}.$$

We define similar expressions for sums of products of tangents of four angles, e.g.,

$$s_2(A, B, C, D) = \tan A \tan B + \tan A \tan C + \tan A \tan D + \tan B \tan C$$
$$+ \tan B \tan D + \tan C \tan D,$$

and of five angles, e.g.,

$$s_4(A, B, C, D, E) = \tan A \tan B \tan C \tan D + \tan A \tan B \tan C \tan E$$
$$+ \tan A \tan B \tan D \tan E + \tan A \tan C \tan D \tan E$$
$$+ \tan B \tan C \tan D \tan E.$$

Now we find a formula for the tangent of the sum of four angles:

$$\tan(A + B + C + D) = \frac{\tan(A + B + C) + \tan D}{1 - \tan(A + B + C) \tan D}$$

$$= \frac{\frac{s_1(A,B,C) - s_3(A,B,C)}{1 - s_2(A,B,C)} + \tan D}{1 - \left(\frac{s_1(A,B,C) - s_3(A,B,C)}{1 - s_2(A,B,C)}\right) \tan D}$$

$$= \frac{s_1(A, B, C) - s_3(A, B, C) + (1 - s_2(A, B, C)) \tan D}{1 - s_2(A, B, C) - (s_1(A, B, C) - s_3(A, B, C)) \tan D}$$

$$= \frac{s_1(A, B, C) - s_3(A, B, C) + \tan D - s_2(A, B, C) \tan D}{1 - s_2(A, B, C) - s_1(A, B, C) \tan D + s_3(A, B, C) \tan D}$$

$$= \frac{s_1(A, B, C, D) - s_3(A, B, C, D)}{1 - s_2(A, B, C, D) + s_4(A, B, C, D)}.$$

Try to take the next step and deduce the following formula for the tangent of the sum of five angles:

$$\tan(A + B + C + D + E) = \frac{s_1(A, B, C, D, E) - s_3(A, B, C, D, E) + s_5(A, B, C, D, E)}{1 - s_2(A, B, C, D, E) + s_4(A, B, C, D, E)}.$$

A pattern is apparent. Given positive integers n and k, such that $1 \leq k \leq n$, let s_k stand for $s_k(\tan \theta_1, \ldots, \tan \theta_n)$, the sum of all products of k of the tangents. Then the formula for the tangent of the sum of n angles is

$$\tan(\theta_1 + \cdots + \theta_n) = \frac{s_1 - s_3 + s_5 - \cdots}{1 - s_2 + s_4 - \cdots},$$

where we set $s_k = 0$ if $k > n$, so that the series in the numerator and denominator are finite.

Let's give an aha! proof of this formula. The proof uses complex numbers and Euler's formula
$$\cos\theta + i\sin\theta = e^{i\theta}.$$
Setting θ equal to the sum of our n angles, we obtain
$$\cos(\theta_1 + \cdots + \theta_n) + i\sin(\theta_1 + \cdots + \theta_n) = e^{i(\theta_1 + \cdots + \theta_n)} = e^{i\theta_1} \ldots e^{i\theta_n},$$
and hence the identity
$$\cos(\theta_1 + \cdots + \theta_n) + i\sin(\theta_1 + \cdots + \theta_n) = (\cos\theta_1 + i\sin\theta_1)\ldots(\cos\theta_n + i\sin\theta_n).$$

The real part of the left side of this identity is $\cos(\theta_1 + \cdots + \theta_n)$. The real part of the right side consists of products of an even number of $i\sin\theta_i$ terms (since $i^2 = -1$) together with a complementary number of $\cos\theta_i$ terms. Hence, each product consists, for some m, of $2m$ terms of the form $\sin\theta_i$ and $n - 2m$ terms of the form $\cos\theta_i$, multiplied by $(-1)^m$. The imaginary part of the left side of the identity is $i\sin(\theta_1 + \cdots + \theta_n)$. The imaginary part of the right side consists of products of an odd number of $i\sin\theta_i$ terms together with a complementary number of $\cos\theta_i$ terms. Hence, each product consists, for some m, of $2m + 1$ terms of the form $\sin\theta_i$ and $n - 2m - 1$ terms of the form $\cos\theta_i$, multiplied by $i(-1)^m$.

Upon dividing both sides of the imaginary part of the identity by the identity for $\cos(\theta_1 + \cdots + \theta_n)$, we obtain on the left side $i\tan(\theta_1 + \cdots + \theta_n)$. On the right side, we also divide numerator and denominator by $\cos\theta_1 \ldots \cos\theta_n$, thus killing off all the $\cos\theta_i$ terms and turning the $\sin\theta_i$ terms into $\tan\theta_i$ terms. The resulting monomials can be grouped together to form all terms of the form $i(-1)^m s_{2m+1}$ in the numerator and all terms of the form $(-1)^m s_{2m}$ in the denominator. Dividing by i establishes the claimed formula.

Bonus: Elementary Symmetric Polynomials

The elementary symmetric polynomials of order n, in the variables $x_1, x_2, \ldots, x_n$, are
$$s_1 = x_1 + x_2 + \cdots + x_n,$$
$$s_2 = \sum_{1 \leq i < j \leq n} x_i x_j,$$
$$\vdots$$
$$s_n = x_1 x_2 \ldots x_n.$$
For example, with $n = 3$, we have
$$s_1 = x_1 + x_2 + x_3,$$
$$s_2 = x_1 x_2 + x_1 x_3 + x_2 x_3,$$
$$s_3 = x_1 x_2 x_3.$$
In our Problem, we worked with $x_i = \tan\theta_i$.

We will see in the Bonus to "Perrin's Sequence" that the elementary symmetric polynomials can be combined to produce any symmetric polynomial.

We show here Newton's identities relating the elementary symmetric polynomials to the symmetric polynomials that are sums of powers of the variables. Let's look at an example. Define

$$p_0 = x_1^0 + x_2^0 + x_3^0 = 3$$
$$p_1 = x_1 + x_2 + x_3$$
$$p_2 = x_1^2 + x_2^2 + x_3^2$$
$$p_3 = x_1^3 + x_2^3 + x_3^3.$$

Then

$$\begin{aligned} p_3 &= x_1^3 + x_2^3 + x_3^3 \\ &= (x_1+x_2+x_3)(x_1^2+x_2^2+x_3^2) - (x_1x_2+x_2x_3+x_3x_1)(x_1+x_2+x_3) + 3x_1x_2x_3 \\ &= s_1 p_2 - s_2 p_1 + s_3 p_0. \end{aligned}$$

In general, with $p_k = x_1^k + x_2^k + \cdots + x_n^k$ and $s_0 = 1$, we have Newton's identities:

$$\sum_{k=0}^{n} s_k p_{n-k} (-1)^k = 0, \quad n \geq 1.$$

1.4 No Calculus Needed

A Zigzag Path

Two points A and B lie on one side of a line l. Construct the shortest path that starts at A, touches l, and ends at B.

• A

• B

l

Solution

We know that a straight line is the shortest distance between two points. Can we use this fact?

As in the figure below, let B' be the symmetric point to B on the other side of l. Given any choice of point P on l, the distance from P to B is the same as the distance from P to B'. Hence, the length of the path from A to P to B is the same as the length of the path from A to P to B'. The shortest path from A to B' is the straight line segment AB'. Therefore, the intersection of this line segment with l determines the point P such that the path from A to l to B is the shortest possible.

1.4 No Calculus Needed

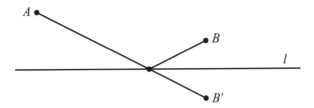

Bonus: Another Zigzag Path

Given two parallel lines l and l', and points A and B between l and l', what is the shortest path that starts at A, touches l, then touches l', and ends at B?

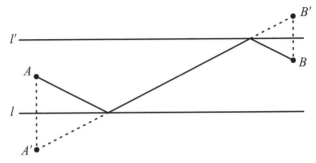

Let A' be the point symmetric to A with respect to l, and B' the point symmetric to B with respect to l'. Construct the line from A' to B'. The intersections of this line with l and l' are the points where the desired path touches the parallel lines.

I. M. Yaglom in the books [21] and [22] presents this and other beautiful geometric construction problems.

A Stack of Circles

What is the sum of the circumferences of the infinite stack of circles inside the triangle in the picture below? (Each circle is tangent to sides of the triangle and to the circles above or below it.)

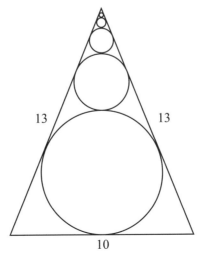

The height of the triangle is 12 (by the Pythagorean theorem). Hence, the sum of the diameters of the circles is 12. The formula for the circumference of a circle of diameter d is $C = \pi d$. Therefore, the sum of the circumferences of all the circles is 12π.

Isn't this easier than finding the diameter of each circle, applying the circumference formula, and summing the infinite series?

Bonus: A Sum of Areas

What is the sum of the areas of the circles in the diagram above? This is easy to find using simple geometry and the sum of a geometric series (see p. 16). Let r be the radius of the largest circle (see the diagram below). Drop altitudes from the center of the largest circle to the three sides of the triangle, thereby dividing the triangle into three smaller triangles. The sum of the areas of these smaller triangles is equal to the area of the given triangle. Hence

$$\frac{1}{2}10r + \frac{1}{2}13r + \frac{1}{2}13r = \frac{1}{2}10 \cdot 12,$$

and $r = 10/3$. (By the way, we'll use this method for finding the inradius of a triangle in "Cutting and Pasting Triangles.")

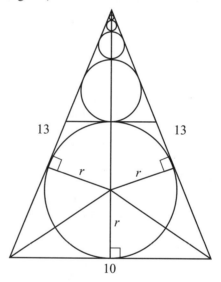

Constructing a tangent line to the largest circle at its top, we see that the given triangle is cut into two pieces, with the top piece similar to the given triangle. The given triangle has height 12. The height of the smaller, similar triangle is $12 - 2r = 16/3$. Hence, the ratio of the smaller height to the larger one is

$$x = \frac{16/3}{12} = \frac{4}{9}.$$

The second-largest circle has radius rx, the third largest rx^2, etc. Therefore, the sum of the areas of all the circles is

$$\pi \left(\frac{10}{3}\right)^2 (1 + x^2 + x^4 + \cdots) = \pi \left(\frac{10}{3}\right)^2 \left(\frac{1}{1-x^2}\right) = \frac{180\pi}{13}.$$

1.4 No Calculus Needed

A Farmer's Field

A farmer wishes to fence in a rectangular field using 300 meters of fence. One side of the field needs no fencing since it is alongside an already existing fence. What is the largest possible area of the field and what dimensions give this area?

Solution

Let the length of fence parallel to the existing fence be x and the two other lengths be y. We are given that $x + 2y = 300$ m.

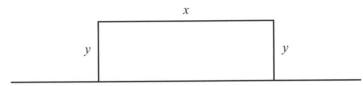

It's a common calculus problem to determine the dimensions x and y that yield the maximum area xy. However, we can avoid calculus by using symmetry. Consider the mirror image of the field with respect to the existing fence.

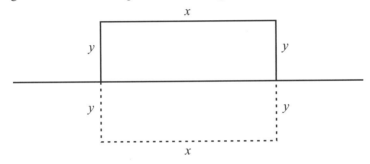

The field and its mirror image together make a rectangle of dimensions $x \times 2y$. Because the field is always half of the area of this larger rectangle, whatever dimensions maximize the area of the larger rectangle also maximize the area of the field. The perimeter of the larger rectangle is fixed (it is 600 m). Since the rectangle of fixed perimeter with largest area is a square, the maximum area of the field occurs when $x = 2y$. Hence, the dimensions that yield the maximum area of the field are $y = 100$ m and $x = 200$ m, and the area is $xy = 20{,}000$ m².

Bonus: A Can of Minimum Surface Area

A cylindrical can is to be made to hold a certain volume. In what proportion are its radius and height so that the can has minimum surface area?

Let h be the height of the can and r the radius of its base. We will show that $h = 2r$. Let the can be inscribed in a rectangular box with dimensions $2r \times 2r \times h$, with its base circumscribed in the base of the box. The ratio of the area of the base of the cylinder to the area of the base of the box is $\pi r^2/(2r)^2 = \pi/4$. The ratio of the area of the top of the cylinder to the area of the top of the box is of course the same. The ratio of the lateral surface area of the cylinder to the lateral surface area of the box is $2\pi rh/8rh = \pi/4$. Hence, the ratio of the total surface area of the can to the total surface area of the box is

$\pi/4$. The ratio of the volume of the can to the volume of the box is $\pi r^2 h/(2r)^2 h = \pi/4$. The fact that these ratios are equal is unimportant; the relevant thing is that they are both constants. Therefore, the surface area of the can is minimized when the surface area of the circumscribing box is minimized. Since a box of given volume is minimized when the box is a cube, the minimum surface area occurs when $h = 2r$. I leave it to you to prove the fact that a box of given volume and minimum surface area is a cube. You can do it with two applications of the isoperimetric inequality that the rectangle of given perimeter having the greatest area is a square (see p. 36).

Composting—A Hot Topic?

Suppose that a compost bin is to be made so that the structure looks from above like the letter E, with three parallel equal-length pieces of fence in one direction and two equal-length pieces of fence in a line in the perpendicular direction. Given that the total length of fencing is fixed, what dimensions yield the maximum area of the bin?

Solution

Let the segments of fence have lengths x and y, as shown in the left diagram below.

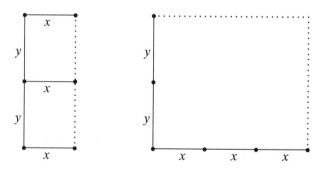

We seek to maximize the area of the rectangle bounded by the E, that is, $x(2y)$, subject to the constraint $3x + 2y = f$, where f is the amount of fencing available. We could solve this problem using calculus, but there is a neat solution using only geometric reasoning. Rearrange the segments of fence according to the right diagram above. The area of this rectangle is $(3x)(2y)$, which is three times the area that we started with. Hence, the values of x and y that maximize the new area are the same as those that maximize the original area. The perimeter of the rectangle is fixed (it is $6x + 4y = 2f$). The area of a rectangle of fixed perimeter is maximized when the rectangle is a square. In our case this means that $3x = 2y = f/2$, and so $x = f/6$ and $y = f/4$. Thus, half of the fencing is used in the set of three parallel pieces and half in the set of two collinear pieces.

By the way, we can give an aha! proof that the square is the rectangle of fixed perimeter with the greatest area. This is called an *isoperimetric inequality*. Suppose that we have a square and a rectangle (not a square) of the same perimeter. We will show that the square has the greater area. As in the diagram below, the square $ABCD$ and the rectangle $AEFG$ are positioned so that they share a vertex A and have parallel corresponding sides. The rectangle cuts the sides of the square at points G and H. The rectangle and square have the shaded area in common. Since the perimeters of the square and rectangle are equal, $|HF| = |BE| = |DG| = |CH|$. It follows that the unshaded area in the square is greater than the unshaded area in the rectangle (since $|CD| > |EF|$). Therefore, the area of $ABCD$ is greater than the area of $AEFG$.

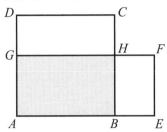

Bonus: The AM--GM Inequality

The arithmetic mean–geometric mean (AM–GM) inequality (see Toolkit) follows instantly. Let $a = |AE|$ and $b = |AG|$. Then

$$ab \leq \left(\frac{a+b}{2}\right)^2,$$

or

$$\sqrt{ab} \leq \frac{a+b}{2}.$$

Equality occurs if and only if $a = b$.

Three Sines

Find the maximum value of $\sin A \sin B \sin C$, where A, B, and C are the angles of a triangle.

Solution

We are given that A, B, and C are positive numbers whose sum is π; hence $\sin A$, $\sin B$, and $\sin C$ are positive. Two steps make finding the maximum value simple. First, by the arithmetic mean–geometric mean inequality,

$$\sin A \sin B \sin C \leq \left(\frac{\sin A + \sin B + \sin C}{3}\right)^3,$$

with equality if and only if $A = B = C = \pi/3$. Second, since $\sin x$ is a concave downward function (see Bonus) for $0 < x < \pi$, we have

$$\frac{\sin A + \sin B + \sin C}{3} \leq \sin\left(\frac{A+B+C}{3}\right),$$

with equality if and only if $A = B = C = \pi/3$. Therefore

$$\sin A \sin B \sin C \leq \sin^3\left(\frac{A+B+C}{3}\right) = \sin^3\frac{\pi}{3} = \left(\frac{\sqrt{3}}{2}\right)^3 = \frac{3\sqrt{3}}{8},$$

with equality if and only if $A = B = C = \pi/3$.

Bonus: Convex Functions

A real-valued function f is *convex* on an interval I if
$$f((1 - \lambda)a + \lambda b) \leq (1 - \lambda)f(a) + \lambda f(b),$$
for all $a, b \in I$ and $0 \leq \lambda \leq 1$. Geometrically speaking, f is convex if f lies below its secants. In the diagram below, which illustrates the case $\lambda = 1/2$, we can see that $f\left(\frac{a+b}{2}\right) \leq \frac{f(a)+f(b)}{2}$.

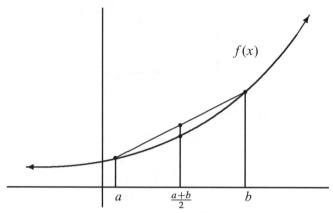

JENSEN'S INEQUALITY. Suppose that f is convex on I. If $a_1, \ldots, a_n \in I$ and $\lambda_1, \ldots, \lambda_n$ are nonnegative real numbers such that $\lambda_1 + \cdots + \lambda_n = 1$, then
$$f\left(\sum_{i=1}^{n} \lambda_i a_i\right) \leq \sum_{i=1}^{n} \lambda_i f(a_i).$$
Equality occurs if and only if all the a_i are equal or f is a linear function.

If the function $-f$ is convex (i.e., f is concave downward), as is the case for the sine function in our Problem, then the inequality is reversed.

Let's prove Jensen's inequality in the case where $n = 4$ and $\lambda = 1/2$. Since f is convex (we'll take the interval I for granted),
$$f\left(\frac{a+b}{2}\right) \leq \frac{f(a) + f(b)}{2},$$
with equality if and only if $a = b$ (we'll assume that f is not a linear function). Now we can use this hypothesis twice:

$$\begin{aligned}
f\left(\frac{a_1 + a_2 + a_3 + a_4}{4}\right) &= f\left(\frac{\frac{a_1+a_2}{2} + \frac{a_3+a_4}{2}}{2}\right) \\
&\leq \frac{f\left(\frac{a_1+a_2}{2}\right) + f\left(\frac{a_3+a_4}{2}\right)}{2} \\
&\leq \frac{\frac{f(a_1)+f(a_2)}{2} + \frac{f(a_3)+f(a_4)}{2}}{2} \\
&= \frac{f(a_1) + f(a_2) + f(a_3) + f(a_4)}{4}.
\end{aligned}$$

1.4 No Calculus Needed

Our inequality is established, with equality if and only if $a_1 = a_2 = a_3 = a_4$.

We can use the same procedure to prove Jensen's inequality for n any power of 2. But how do we prove it for, say $n = 3$? We use a little technique credited to Augustin Louis Cauchy (1789–1857). Set $a_4 = (a_1 + a_2 + a_3)/3$. Then $(a_1 + a_2 + a_3 + a_4)/4 = (a_1 + a_2 + a_3)/3$, and hence

$$f\left(\frac{a_1 + a_2 + a_3}{3}\right) \leq \frac{f(a_1) + f(a_2) + f(a_3) + f\left(\frac{a_1+a_2+a_3}{3}\right)}{4}.$$

Simplifying, we obtain

$$f\left(\frac{a_1 + a_2 + a_3}{3}\right) \leq \frac{f(a_1) + f(a_2) + f(a_3)}{3}.$$

Equality occurs if and only if $a_1 = a_2 = a_3$. The general statement of Jensen's inequality can be proved similarly or by induction on n.

2
Intermediate Problems

I hope that you enjoyed the elementary problems. Now let's try a selection of somewhat more difficult problems. In this chapter, we can expect to use familiar techniques from calculus and other branches of mathematics. As usual, we are looking for illuminating proofs. Aha! solutions are to be found!

2.1 Algebra

Passing Time

At some time between 3:00 and 4:00, the minute hand of a clock passes the hour hand. Exactly what time is this? (Assume that the hands move at uniform rates.)

Solution

Let's solve the problem in a mundane way first (before giving an aha! solution). We reckon time in minutes from the top of the hour (12 on the clock). Suppose that the minute hand is at t minutes and the hour hand is, correspondingly, at $15 + t/12$ minutes. (The term $t/12$ is due to the fact that in 60 minutes the hour hand advances 5 minutes.) When the two hands coincide, we have $t = 15 + t/12$, and hence $t = 16 + 4/11$. Therefore the solution is 3:16+4/11 minutes.

In order to delve deeper into this problem, let's answer the same question for the time between 2:00 and 3:00 when the minute hand and hour hand coincide. Using the same reasoning as before, we obtain $t = 10 + t/12$, and hence $t = 10 + 10/11$, yielding the solution 2:10+10/11 minutes. We see fractions with 11 in the denominator in both cases. Reflecting on this, we realize that in the course of twelve hours the minute hand and hour hand coincide eleven times. Specifically, this happens at 12:00; at some time between 1:00 and 2:00; between 2:00 and 3:00 (we already figured this one out); between 3:00 and 4:00 (we figured this one out, too); between 4:00 and 5:00; between 5:00 and 6:00; between

41

6:00 and 7:00; between 7:00 and 8:00; between 8:00 and 9:00; between 9:00 and 10:00; and between 10:00 and 11:00. The two hands do not coincide between 12:00 and 1:00, or between 11:00 and 12:00, because the minute hand travels faster than the hour hand. By symmetry, the eleven coincidences are equally spaced around the circular face of the clock.

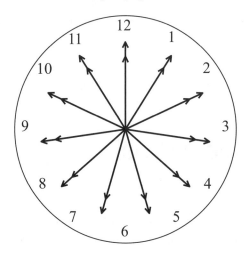

We immediately see that the coincidence between 3:00 and 4:00 occurs 3/11 of the way around the circle, i.e., at $3/11 \times 60 = 16 + 4/11$ minutes past 3:00.

Bonus: What About the Second Hand?

There is no time other than at 12:00 when the hour hand, the minute hand, and the second hand of the (ideal) clock all coincide. The reason is that the minute hand and second hand coincide at 59 points equally spaced around the circle, and the numbers 11 and 59 have no common factor. (By the way, the hour hand and second hand coincide at 719 points equally spaced around the circle.) Other than near 12:00, when are a clock's second hand, minute hand, and hour hands closest together? (This problem is posed and answered in Martin Gardner's treasure trove [7] but the solution isn't explained in detail.)

The strategy is to look at the eleven coincidences between minute hand and hour hand, and determine where the second hand is at those times. When the second hand is closest, that will give us nearly the time at which the second, minute, and hour hands are closest.

The coincidence of minute hand and hour hand between 1:00 and 2:00 occurs 1/11 around the circle, which is at $5 + 5/11$ minutes, and hence the second hand is 5/11 around the circle (rather far from the coincidence of minute hand and hour hand). Similarly, the coincidence of minute hand and hour hand between 2:00 and 3:00 occurs 2/11 around the circle, which is at $10 + 10/11$ minutes, and hence the second hand is 10/11 around the circle (also far). Continuing in this manner, we find that the coincidences of minute hand and hour hand at which the second hand is closest are at 3:16+4/11 minutes and 8:43+7/11 minutes. At these times, the second hand is 4/11 and 7/11 of the way around the circle, respectively. In both cases, when the second hand gets as close as possible, it coincides with the hour hand (since the minute hand moves faster than the hour hand). The coincidence between second hand and hour hand (measured in seconds) that occurs between 3:16 and

3:17 is given by
$$t = 15 + \frac{16}{12} + \frac{t}{720},$$
and hence $t = 11760/719$ (seconds) $= 16 + 196/719$ (minutes). So the solution is 3:16:16+256/719 seconds. The difference between the minute hand and hour hand at this time is
$$\frac{16 + 256/719}{60} - \frac{16 + 196/719}{60} = \frac{1}{719} \text{ of the circle.}$$
This makes sense because the minute hand is at some multiple of $1/719$ around the circle (as its position is dictated by the hour hand's position), and the closest it can be is $1/719$ of the circle. The other (symmetric) solution is 8:43:43+463/719 seconds.

Sums to 1,000,000

Find all sequences of consecutive positive integers that sum to 1,000,000.

Solution

This is the kind of problem that we could crunch out on a computer, but let's try to find a deft mathematical solution.

The crux of the method lies in what to focus on. Let n be the number of terms in the sequence and v the average of the terms. Then
$$nv = 1,000,000.$$
If v is an integer, then both n and v are integer divisors of $1,000,000$. When is v an integer? If n is odd, then v is an integer, since it is equal to the middle term of the sequence. However, if n is even, then v is not an integer; it is equal to the average of the two middle terms of the sequence (and hence is a half-integer). So it seems that we must consider the cases n odd and n even separately, and that the n even case is a little tricky. Believe it or not, there is a way to ensure that n is odd, and this is where the aha! realization comes in.

The problem asks for sequences of positive integers but let's relax the condition temporarily to any integers (positive, negative, or zero). If we have a sum
$$a + \cdots + b$$
of consecutive positive integers, then
$$-(a-1) + \cdots + (a-1) + a + \cdots + b$$
is also a sum of consecutive integers. The new sum is equal to the old sum, as we have introduced only 0 (if $a = 1$) or 0 and pairs of numbers that sum to 0 (if $a > 1$). Moreover, the parity of the number of terms switches (as we have introduced an odd number of new terms). Therefore, the sequences we are looking for occur in companion pairs, one with an even number of terms and one with an odd number of terms, and one containing only positive integers and the other containing some non-positive integers.

By the above discussion of the companion sequences, we may focus on the sequences where n is odd and therefore v is an integer; that is to say, n is an odd divisor of 1,000,000.

Since the prime factorization of 1,000,000 is $2^6 \cdot 5^6$, the choices for n are 1, 5, 5^2, 5^3, 5^4, 5^5, and 5^6, and each of these seven choices gives rise to exactly one solution (the resulting sequence or its companion). The respective values of v are 1,000,000, 200,000, 40,000, 8000, 1600, 320, and 64. Hence, the sums and their companion sums are, respectively,

1,000,000	$-999{,}999 + \cdots + 1{,}000{,}000$
$199{,}998 + \cdots + 200{,}002$	$-199{,}997 + \cdots + 200{,}002$
$39{,}988 + \cdots + 40{,}012$	$-39{,}987 + \cdots + 40{,}012$
$7938 + \cdots + 8062$	$-7937 + \cdots + 8062$
$1288 + \cdots + 1912$	$-1287 + \cdots + 1912$
$-1242 + \cdots + 1882$	$1243 + \cdots + 1882$
$-7748 + \cdots + 7876$	$7749 + \cdots + 7876.$

For example, to find the sum for $n = 5$ and $v = 200{,}000$, we calculate as the beginning and ending terms $v - (n-1)/2 = 199{,}998$ and $v + (n-1)/2 = 200{,}002$.

We confirm that seven sequences have all positive terms.

Bonus: Sums to Any Number

For the general case of sequences of consecutive positive integers that sum to any given "target integer," the argument is similar to the one above. The number of such sequences is the number of odd divisors of the target integer. Hence, it is the product of terms of the form $e + 1$, where p^e ranges over all odd prime power divisors of the target. If the target integer is a power of 2, then the only such sequence is the number itself. (See "fundamental theorem of arithmetic" in the Toolkit.)

An Odd Determinant

Does the matrix

$$\begin{bmatrix} 10^{30} + 5 & 10^{10} + 4 & 10^7 + 2 & 10^{18} + 10 \\ 10^5 + 6 & 10^{100} + 3 & 10^8 + 10 & 10^{15} + 20 \\ 10^{14} + 80 & 10^{19} + 4 & 10^4 + 5 & 10^{40} + 4 \\ 10^{50} + 2 & 10^{13} + 6 & 10^{23} + 8 & 10^9 + 17 \end{bmatrix}$$

have a multiplicative inverse?

Solution

Let's determine whether the determinant of this matrix is even or odd. Replacing all even entries by 0 and all odd entries by 1, we obtain the matrix

$$\begin{bmatrix} 1 & 0 & 0 & 0 \\ 0 & 1 & 0 & 0 \\ 0 & 0 & 1 & 0 \\ 0 & 0 & 0 & 1 \end{bmatrix}.$$

2.1 Algebra

The determinant of this matrix is odd (it is 1). So the same is true of the original matrix (since the determinant of a matrix is computed by addition and multiplication operations on its entries). Hence, the determinant of the matrix is not 0 and the matrix is invertible.

Bonus: Inside or Outside?

Here is another problem that looks baffling at first but has a simple parity (even/odd) solution. Can you tell at a glance whether the point is inside or outside the curve?

An easy way to do it is to draw a ray from the point and count the number of times it crosses the curve. Each time the ray crosses the curve, it moves from inside the curve to outside the curve or from outside the curve to inside the curve (this follows from the Jordan curve theorem). In the picture below, the ray crosses the curve eleven times. The odd number of crossings means that the point is inside the curve.

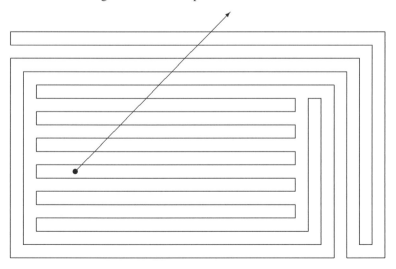

Demanding a Polynomial

Is there a polynomial $p(x)$ with integer coefficients such that $p(1) = 3$ and $p(3) = 2$?

Solution

There is no such polynomial. For if we consider all operations modulo 2, the two requirements are contradictory: $p(1) \equiv 1 \pmod{2}$ and $p(1) \equiv 0 \pmod{2}$.

Bonus: Lagrange's Interpolation Formula

Such a polynomial does exist—in fact, a linear one—with *rational* coefficients. Suppose that $p(x) = mx + b$. Then $m = (2-3)/(3-1) = -1/2$, and the point–slope formula yields $p(x) - 3 = -1/2(x - 1)$, or $p(x) = -1/2x + 7/2$.

Here is a general recipe for creating polynomials with prescribed values.

LAGRANGE'S INTERPOLATION FORMULA.[1] Let $\alpha_0, \alpha_1, \ldots, \alpha_n$ be distinct complex numbers and $\beta_0, \beta_1, \ldots, \beta_n$ arbitrary complex numbers.

(a) The polynomial

$$P(z) = \sum_{i=0}^{n} \beta_i \prod_{\substack{0 \le j \le n \\ j \ne i}} \frac{z - \alpha_j}{\alpha_i - \alpha_j}$$

has the property that $P(\alpha_i) = \beta_i$, for $i = 0, 1, \ldots, n$.

(b) There is exactly one polynomial of degree at most n with the property specified in (a).

Statement (a) is verified by evaluating P at $\alpha_0, \alpha_1, \ldots, \alpha_n$ and noting that in each case all terms in the sum but one are 0, yielding $P(\alpha_i) = \beta_i$.

To prove (b), suppose that P and Q are two such polynomials. Let $R(z) = P(z) - Q(z)$. Then R has degree at most n and has $n + 1$ roots, namely, $\alpha_0, \alpha_1, \ldots, \alpha_n$. Hence, R is identically 0, and P and Q are identical polynomials. This proves uniqueness.

2.2 Geometry

What's the Side Length?

Let P be a point inside an equilateral triangle ABC, with $PA = 4$, $PB = 3$, and $PC = 5$. Find x, the side length of the triangle.

---

[1]This formula is credited to Joseph-Louis Lagrange (1736–1813). Lagrange made contributions in analysis, number theory, and classical and celestial mechanics.

2.2 Geometry

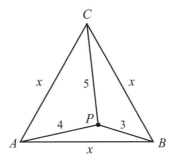

Solution

As in the figure below, rotate $\triangle ABC$ 60° clockwise around B. Suppose that P and C are transformed to P' and C', respectively. We see that $\triangle PP'B$ is equilateral and hence $PP' = 3$. Since $\triangle PP'C$ has side lengths 3, 4, and 5, it follows that $\angle PP'C$ is a right angle. Hence

$$m(\angle BP'C) = 60° + 90° = 150°.$$

By the law of cosines (see Toolkit),

$$x^2 = 3^2 + 4^2 - 2 \cdot 3 \cdot 4 \cdot \cos 150° = 25 + 12\sqrt{3}.$$

Therefore

$$x = \sqrt{25 + 12\sqrt{3}}.$$

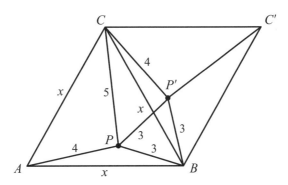

This innocent problem is a gateway to a lot of fascinating mathematics. We'll see some in the Bonus.

Bonus: Integer Solutions

Let the distances from P to the three vertices be a, b, and c, as in the diagram below. With similar reasoning to that in the Solution, we can derive a formula for x.

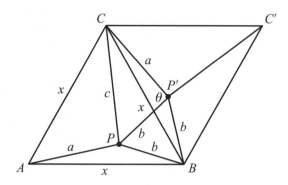

We have
$$x^2 = a^2 + b^2 - 2ab\cos(60° + \theta),$$
where
$$c^2 = a^2 + b^2 - 2ab\cos\theta.$$

Eliminating θ, we obtain

$$x = \sqrt{\frac{a^2 + b^2 + c^2}{2} \pm \frac{\sqrt{3}}{2}\sqrt{2a^2b^2 + 2a^2c^2 + 2b^2c^2 - a^4 - b^4 - c^4}}.$$

The positive sign is the one relevant to our problem. The negative sign gives the solution if P is a point in the exterior of the triangle. Because a, b, and c are the side lengths of $\triangle PP'C$, they satisfy the weak triangle inequality $a + b \leq c$, where we take c to be the largest of the three distances.

Rearranging our equation for x yields

$$x^4 + a^4 + b^4 + c^4 = x^2a^2 + x^2b^2 + x^2c^2 + a^2b^2 + a^2c^2 + b^2c^2.$$

Is it possible to find an equilateral triangle with integer side lengths and a point P in the interior of the triangle such that the distances from P to the three vertices are all integers? Yes, a computer search reveals that the smallest integer solution is $a = 57, b = 65, c = 73$, and $x = 112$.

Using Ptolemy's theorem (see Toolkit), one can find infinitely many integer solutions in which P is on the circumcircle of the triangle. One such solution is $\{a, b, c, x\} = \{3, 5, 8, 7\}$. Similarly, one can find infinitely many integer solutions in which P is on an extended side of the triangle. Let's call solutions trivial in which P is on the circumcircle of the triangle or on a line determined by a side of the triangle. In 1990 Arnfried Kemnitz found an infinite family of nontrivial integer solutions in which P is in the plane of the triangle:

$$a = m^2 + n^2$$
$$b = m^2 - mn + n^2$$
$$c = m^2 + mn + n^2$$
$$m = 2(u^2 - v^2)$$
$$n = u^2 + 4uv + v^2$$
$$x = 8(u^2 - v^2)(u^2 + uv + v^2),$$

where u and v are any positive integers with $u > v$. Notice that in this infinite family we always have $b + c = 2a$. The computer solution mentioned above doesn't have this property, so not all nontrivial solutions are in Kemnitz' family. In fact, the quartic surface in question cannot be parameterized (it is an *irrational surface*). The surface, called a tetrahedroid, is a special kind of quartic known as a Kummer surface, named after number theorist and algebraic geometer Ernst Eduard Kummer (1810–1893). A three-variable version of the surface, via *Mathematica*®, is depicted below.

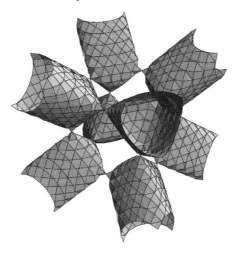

Napoleon's Theorem

Prove:

NAPOLEON'S THEOREM.[2] Given $\triangle ABC$, let A', B', C' be the third vertices of equilateral triangles constructed outwardly on sides BC, CA, AB, respectively, and let A'', B'', C'' be the centers of $\triangle A'BC$, $\triangle B'CA$, $C'AB$, respectively. Then $\triangle A''B''C''$ is equilateral. (See the picture below.)

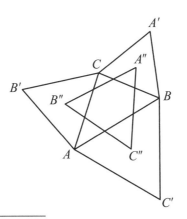

---

[2] It is not known whether Napoleon Bonaparte (1769–1821) was really the first person to discover and prove this theorem, although he is known to have been an able mathematician.

Solution

We give a proof that is easy to "see." Consider the "wallpaper pattern" in the following figure, where the given triangle is black. Look at the hexagon formed by joining the centers of six equilateral triangles that "go around" one of the equilateral triangles (the "focal triangle") constructed on the sides of the given triangle. By symmetry, the sides of this hexagon are all equal. Also by symmetry, alternate "spokes" of the hexagon are equal. (The "'spokes" join the center of the focal triangle to the vertices of the hexagon.) Hence, two consecutive triangles around the center of the hexagon are congruent. Therefore, all six such triangles are congruent, and hence all central angles are $\pi/3$. Since we could have chosen any of the three equilateral triangles about the given triangle to be the focal triangle, the triangle of Napoleon's theorem is equilateral.

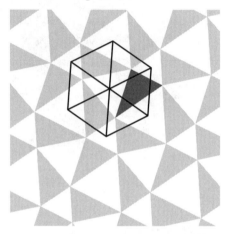

Bonus: An Algebraic Proof

Suppose that A, B, C (given, without loss of generality, in counter-clockwise order) are represented by vectors a, b, c, respectively. Then the third points of the equilateral triangles, A', B', C', are given by a', b', c', respectively, where

$$M(a - b) = c' - b,$$
$$M(b' - c) = a - c,$$
$$M(c - a') = b - a',$$

and M is a rotation matrix around the origin of angle $\pi/3$. (We don't need to write M explicitly.) The centers of the equilateral triangles, A'', B'', C'', are given by a'', b'', c'', respectively, where

$$a'' = \frac{1}{3}(a' + b + c),$$
$$b'' = \frac{1}{3}(a + b' + c),$$
$$c'' = \frac{1}{3}(a + b + c').$$

2.2 Geometry

To show that A'', B'', C'' are the vertices of an equilateral triangle, we will demonstrate that $M(b'' - a'') = c'' - a''$. Here is the algebra:

$$M(b'' - a'') = \frac{1}{3}M(a + b' + c - a' - b - c)$$

$$= \frac{1}{3}M(a - b + b' - c + c - a')$$

$$= \frac{1}{3}(c' - b + a - c + b - a')$$

$$= \frac{1}{3}(a + b + c' - a' - b - c)$$

$$= c'' - a''.$$

By the way, the result in Napoleon's theorem also holds if the equilateral triangles are constructed inwardly rather than outwardly. The area of the "outward Napoleon triangle" is

$$\frac{\sqrt{3}(a^2 + b^2 + c^2)}{24} + \frac{1}{2}A$$

and the area of the "inward Napoleon triangle" is

$$\frac{\sqrt{3}(a^2 + b^2 + c^2)}{24} - \frac{1}{2}A,$$

where A is the area of the given triangle.

Liang-shin Hahn's book [8] gives an elegant proof of Napoleon's theorem via complex numbers.

A Graph on a Doughnut

The complete bipartite graph $K_{4,4}$ consists of two sets of four vertices and all possible edges between the two sets (see Toolkit).

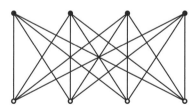

Show that $K_{4,4}$ can be drawn on the surface of a doughnut (torus) without edge-crossings.

Solution

The proof is by a simple picture (below). The square represents the torus. The top edge and bottom edge of the square are "identified," i.e., they are regarded as the same. This identification wraps the square around to make a cylinder. The left edge and right edge of the square are also identified. This glues the two ends of the cylinder together to form a torus. The figure can be stretched to make the gluing happen.

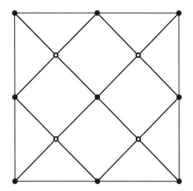

The two sets of vertices in $K_{4,4}$ are represented by solid dots (for one set) and open dots (for the other set). Vertices on the boundary of the square are shown multiple times (because of the identifications of the edges). For instance, the four vertices at the corners of the square are all the same vertex. Our diagram does what we want it to do, since each vertex is joined to the four vertices of the other set and there are no edge crossings.

Here is a Euclidean depiction of the graph on the torus. The other four vertices are on the opposite side of the torus.

Bonus: The Genus of a Graph

We say that a sphere has genus 0. A surface obtained from a sphere by attaching g "handles" is called a *surface of genus g*. For example, a torus is a surface of genus 1. The genus of a graph G is the minimum g such that G can be drawn without edge-crossings on a surface of genus g. We found in the Solution that the complete bipartite graph $K_{4,4}$ has genus at most 1; in fact, it has genus 1. The genus of the complete bipartite graph $K_{m,n}$ is given by the formula

$$\left\lceil \frac{(m-2)(n-2)}{4} \right\rceil,$$

where $\lceil x \rceil$ is the least integer greater than or equal to x. See, for example, [9].

Points Around an Ellipse

Given an ellipse, what is the locus of points P in the plane of the ellipse such that there are two perpendicular tangents from P to the ellipse?

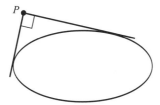

Solution

Let the ellipse be given by
$$\frac{x^2}{a^2} + \frac{y^2}{b^2} = 1.$$

We will prove that the locus is a circle centered at the origin with radius $\sqrt{a^2 + b^2}$.

Let $y = mx + n$ be a tangent to the ellipse. Then the equation
$$\frac{x^2}{a^2} + \frac{(mx+n)^2}{b^2} = 1$$

has a double root in x. We rewrite the equation as
$$\left(\frac{b^2}{a^2} + m^2\right) x^2 + 2mnx + n^2 - b^2 = 0.$$

Since this equation has a double root, its discriminant is 0, i.e.,
$$4m^2n^2 - 4\left(\frac{b^2}{a^2} + m^2\right)(n^2 - b^2) = 0.$$

This equation simplifies to
$$a^2m^2 - n^2 + b^2 = 0.$$

Let P have coordinates (x_0, y_0). Because the tangent lines pass through P, we have $n = y_0 - mx_0$ and hence
$$(a^2 - x_0^2)m^2 + 2x_0y_0 m + b^2 - y_0^2 = 0.$$

Since the two tangents from P to the ellipse are perpendicular, the two roots m_1 and m_2 to this latter equation satisfy $m_1 m_2 = -1$, and it follows that
$$x_0^2 + y_0^2 = a^2 + b^2,$$

which is the equation of a circle centered at the origin with radius $\sqrt{a^2 + b^2}$.

Bonus: The Problem for Hyperbolas and Parabolas

We can ask the same question for a hyperbola and a parabola. As an exercise, use the method of the Solution to prove the following propositions:

(1) The locus of points $P(x_0, y_0)$ with the property that there are two perpendicular tangents from P to the hyperbola $x^2/a^2 - y^2/b^2 = 1$, where $a > b$, is the circle $x_0^2 + y_0^2 = a^2 - b^2$.

(2) The locus of points $P(x_0, y_0)$ with the property that there are two perpendicular tangents from P to the parabola $y = kx^2$ is the horizontal line $y_0 = -1/(4k)$.

Three Fixed Points

A transformation of Euclidean space that preserves distances between points is called an *isometry*. Prove that if an isometry of the Euclidean plane fixes three noncollinear points (i.e., these three points don't move), then the isometry fixes every point in the plane.

Solution

Let A, B, and C be fixed points of an isometry. We will prove the contrapositive form of the assertion: if some point is moved by the transformation, then A, B, C are collinear. Suppose that the transformation moves the point P to a different point P'. Let l be the perpendicular bisector of the line segment joining P and P'. Then A lies on the line l, since the distance from A to P is the same as the distance from A to P'. By the same reasoning, B and C also lie on l. Hence, A, B, and C are collinear.

Bonus: Isometries of the Plane

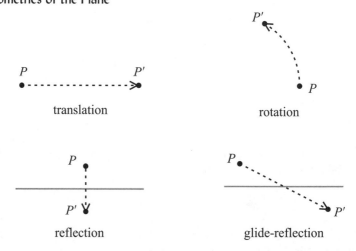

As shown in the diagrams above, the four types of isometries of the plane are translations, rotations, reflections, and glide-reflections. A glide-reflection is a composition of a translation and a reflection in a line parallel to the direction of translation. A translation and a glide-reflection have no fixed points (although a glide-reflection's reflecting line is fixed

2.2 Geometry

as a set of points), while a rotation has one fixed point (the center) and a reflection has a line of fixed points. Translations and rotations are orientation-preserving, while reflections and glide-reflections are orientation-reversing.

Let's show that these are the only types of isometries of the plane.

CLAIM: If f is an isometry of $\mathbf{R}^2$, then

$$f(x) = g(x) + h,$$

where g is an orthogonal linear transformation and $h \in \mathbf{R}^2$.

We will see what an orthogonal linear transformation is presently. Let f be an isometry. Then $g(x) = f(x) - f(0)$ is an isometry, because, for $x, y \in \mathbf{R}^2$, we have

$$\|g(x) - g(y)\| = \|f(x) - f(0) - f(y) + f(0)\| = \|f(x) - f(y)\| = \|x - y\|.$$

(We are denoting the length of a vector v by $\|v\|$.) Furthermore, $g(0) = 0$.

We now show that g preserves the dot product of vectors. For any $x, y \in \mathbf{R}^2$, we have

$$\|g(x) - g(y)\| = \|x - y\|,$$
$$\|g(x) - g(y)\|^2 = \|x - y\|^2,$$
$$(g(x) - g(y)) \cdot (g(x) - g(y)) = (x - y) \cdot (x - y),$$
$$g(x) \cdot g(x) - 2g(x) \cdot g(y) + g(y) \cdot g(y) = x \cdot x - 2x \cdot y + y \cdot y,$$
$$\|g(x)\|^2 - 2g(x) \cdot g(y) + \|g(y)\|^2 = \|x\|^2 - 2x \cdot y + \|y\|^2.$$

Since $\|g(x)\| = \|x\|$ and $\|g(y)\| = \|y\|$, it follows that

$$g(x) \cdot g(y) = x \cdot y.$$

Hence, g preserves the dot product of vectors.

Now suppose that $\{e_1, e_2\}$ is a standard basis for $\mathbf{R}^2$. Then, since g is an isometry, $\|g(e_1)\| = \|g(e_2)\| = 1$. Also,

$$g(e_1) \cdot g(e_2) = e_1 \cdot e_2 = 0.$$

Hence, $\{g(e_1), g(e_2)\}$ is an orthonormal basis for $\mathbf{R}^2$. Furthermore, for

$$x = \lambda_1 e_1 + \lambda_2 e_2,$$

we have

$$g(x) \cdot g(e_i) = x \cdot e_i = \lambda_i, \quad i = 1, 2,$$

and hence

$$g(x) = \lambda_1 g(e_1) + \lambda_2 g(e_2).$$

This is what we mean by g being an orthogonal linear transformation: each vector is mapped to a vector with the same coordinates with respect to the transformed basis. The conclusion of the claim follows with $h = f(0)$.

Now we know that $f(x) = g(x) + h$, where g is given by an orthogonal matrix and h is a vector. What are the possible 2×2 orthogonal matrices? Since the columns of the matrix are orthogonal unit vectors, we may take the first column to be

$$\begin{bmatrix} \cos\theta \\ \sin\theta \end{bmatrix}, \quad 0 \le \theta < 2\pi.$$

Two choices for the second column are

$$\begin{bmatrix} -\sin\theta \\ \cos\theta \end{bmatrix} \quad \text{and} \quad \begin{bmatrix} \sin\theta \\ -\cos\theta \end{bmatrix}.$$

Since there are only two unit vectors orthogonal to the first column, these must be the two! These choices yield two possible matrices:

$$\begin{bmatrix} \cos\theta & -\sin\theta \\ \sin\theta & \cos\theta \end{bmatrix} \quad \text{and} \quad \begin{bmatrix} \cos\theta & \sin\theta \\ \sin\theta & -\cos\theta \end{bmatrix}.$$

The first matrix corresponds to a rotation, $g(x) = R_\theta x$, the second matrix to a reflection in the line through the origin with slope $\theta/2$.

Finally, we should show that upon adding a translation, we obtain the four types of isometries claimed. If $h = 0$, then an isometry of the first type is a (counter-clockwise) rotation around the origin by angle θ. If $h \ne 0$, then such an isometry is a translation (if $\theta = 0$), or otherwise a rotation by θ with center $c = h(I - R_\theta)^{-1}$, since this value of c satisfies the equation

$$R_\theta(x - c) + c = R_\theta x + h.$$

If $h = 0$, then an isometry of the second type is a reflection in the line through the origin with slope $\theta/2$. If $h \ne 0$, then such an isometry is a glide-reflection by h if h is parallel to the line through the origin with slope $\theta/2$. If h is not parallel to this line, then the isometry is a reflection with respect to this line translated by the vector $h/2$. For an explanation of the significance of the half-angle $\theta/2$, see "Reflections and Rotations."

'Round and 'Round

Suppose that two circles of radii r and R intersect at two points. See the picture below.

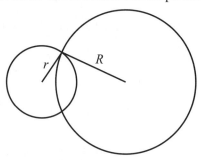

Two particles start at one of the intersection points and move counter-clockwise around the two circles, at constant speeds, completing a revolution of their respective circles at the same time. What is the locus of the midpoint of the line segment joining the two particles?

2.2 Geometry

Solution

Let c_1 be the center of the circle of radius r and c_2 the center of the circle of radius R. Let $p_1(t)$ and $p_2(t)$ be the positions of the two particles at time t. Set $p(0) = p_1(0) = p_2(0)$, i.e., the intersection point where the particles start. Then

$$p_1(t) = M_t(p(0) - c_1) + c_1$$

and

$$p_2(t) = M_t(p(0) - c_2) + c_2,$$

where M_t is a rotation operator corresponding to time t. (We don't need to write M_t explicitly.) Hence, the midpoint $\overline{p}(t)$ of the line segment joining $p_1(t)$ and $p_2(t)$ is given by

$$\overline{p}(t) = \frac{p_1(t) + p_2(t)}{2} = M_t\left(p(0) - \frac{c_1 + c_2}{2}\right) + \frac{c_1 + c_2}{2}.$$

Therefore, the locus is a circle centered at the midpoint of the line segment joining the circles' centers and passing through the circles' two intersection points. See the following diagram.

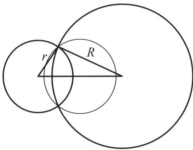

Bonus: A Fundamental Parallelogram

Let x be the radius of the locus circle and $2y$ be the distance between the centers of the two given circles. Then we have a parallelogram with sides r and R, and diagonals $2x$ and $2y$, as shown in the diagram below. As in "The Case of the Rotating Parallelogram," it follows by the Law of Cosines (see Toolkit) that the four parameters satisfy the relation (known as the parallelogram law)

$$r^2 + R^2 = 2x^2 + 2y^2.$$

From this relation, we could find x, the radius of the locus circle, in terms of r, R, and y.

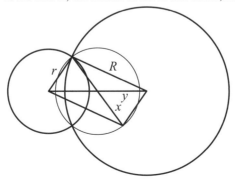

Reflections and Rotations

(a) Suppose that α and β are two intersecting lines in the Euclidean plane. What kind of transformation of the plane results if a reflection is made with respect to α and then a reflection with respect to β?

(b) Suppose that P and P' are two distinct points in the Euclidean plane. What kind of transformation of the plane results if a rotation is made with center P and angle θ and then a rotation with center P' and angle θ', where $\theta + \theta'$ is not an integer multiple of 2π? (The angles are measured counterclockwise.)

Solution

Consider the diagrams below.

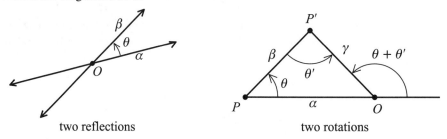

two reflections   two rotations

(a) The figure on the left shows the intersecting lines of reflection, α and β. Let's also call the reflections themselves α and β. The angle between α and β (measured counterclockwise) is θ. We will show that the composition of α and β is a rotation with center O (the intersection of α and β) and angle 2θ.

We use conjugations, expressions of the form gxg^{-1} which perform an operation x with respect to a reference frame given by g. Let r_ω denote a rotation with center O with angle ω. The composition of α and β is

$$\alpha\beta = \alpha(r_{-\theta}\alpha r_\theta) = (\alpha r_{-\theta}\alpha)r_\theta = r_\theta r_\theta = r_{2\theta}.$$

(b) Let's say that the rotation at P has angle 2θ and the rotation at P' has angle $2\theta'$. From (a), we know that the rotation at P is the product of two reflections, α and β, intersecting at an angle θ, as in the figure on the right. (We have taken β to be the line through P and P'.) Similarly, the rotation at P' is the product of two reflections, β and γ, intersecting at an angle θ'. The composition of the two rotations is

$$(\alpha\beta)(\beta\gamma) = \alpha(\beta\beta)\gamma = \alpha\gamma.$$

This is a rotation with center O (the intersection of α and γ) and angle $2(\theta + \theta')$, since the angle between α and β is $\theta + \theta'$.

Bonus: Tiling With Triangles

Which Euclidean triangles have the property that reflections in their sides produce tilings of the Euclidean plane? (Think of flipping a triangle over with respect to its sides.) We will show that there are exactly four such triangles: a 30°–30°–120° triangle, a 30°–60°–90° triangle, a 45°–45°–90° triangle, and a 60°–60°–60° triangle.

2.2 Geometry

Since in the tiling the triangle is rotated around each vertex, the angles of the triangle are $2\pi/a, 2\pi/b, 2\pi/c$, where a, b, c are integers greater than 2, and $2\pi/a + 2\pi/b + 2\pi/c = \pi$. Hence

$$\frac{1}{a} + \frac{1}{b} + \frac{1}{c} = \frac{1}{2}.$$

Suppose that $3 \leq a \leq b \leq c$. We see that $3 \leq a \leq 6$, for if $a > 6$ then the sum of the angles is too small to make a triangle. If $a = 6$, then $(a, b, c) = (6, 6, 6)$ is the only solution. If $a = 5$, then $c = 10b/(3b - 10)$, and we find that the only solution is $(a, b, c) = (5, 5, 10)$. Continuing in this manner, we determine all the solutions: $(a, b, c) = (3, 7, 42), (3, 8, 24), (3, 9, 18), (3, 10, 15), (3, 12, 12), (4, 5, 20), (4, 6, 12), (4, 8, 8), (5, 5, 10), (6, 6, 6)$.

By inspection (you may want to check this), only the triples $(3, 12, 12)$, $(4, 6, 12)$, $(4, 8, 8)$, and $(6, 6, 6)$ yield tilings. (The other triples correspond to triangles that overlap each other when reflected.) The corresponding triangles have angle measures 30°–30°–120°, 30°–60°–90°, 45°–45°–90°, and 60°–60°–60°, respectively. The four triangle tilings are illustrated below.

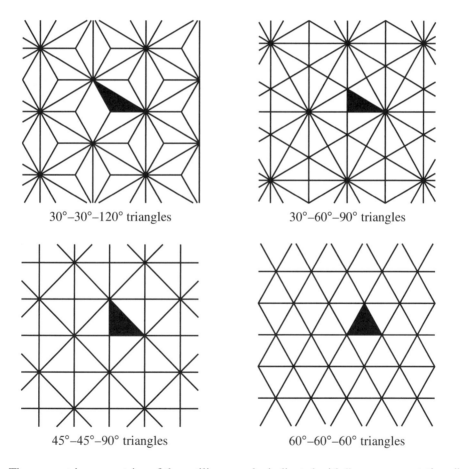

30°–30°–120° triangles   30°–60°–90° triangles

45°–45°–90° triangles   60°–60°–60° triangles

The geometric symmetries of these tilings can be indicated with "group presentations." (See "Bonus: A Tiling of the Hyperbolic Plane.") The four group presentations, corresponding to the four tilings, are given below.

tiling	group presentation
30°–30°–120° triangles	$\langle a, b, c : a^2 = b^2 = c^2 = (ab)^3 = (bc)^6 = (ca)^6 = 1 \rangle$
30°–60°–90° triangles	$\langle a, b, c : a^2 = b^2 = c^2 = (ab)^2 = (bc)^6 = (ca)^3 = 1 \rangle$
45°–45°–90° triangles	$\langle a, b, c : a^2 = b^2 = c^2 = (ab)^2 = (bc)^4 = (ca)^4 = 1 \rangle$
60°–60°–60° triangles	$\langle a, b, c : a^2 = b^2 = c^2 = (ab)^3 = (bc)^3 = (ca)^3 = 1 \rangle$

In each group presentation, the "generators" a, b, c represent reflections with respect to the three sides of the given triangle. That is why we have the relations $a^2 = 1$, $b^2 = 1$, and $c^2 = 1$ (two reflections with respect to the same line yield the identity). As we know from the Solution, the composition of two reflections yields a rotation by an angle twice the angle between the two reflecting lines. For example, in the tiling with 60°–60°–60° triangles, the product ab is a rotation by 120°. This is why we have the relation $(ab)^3 = 1$ (three of these rotations yield the identity). A "word" in the group is any sequence of a's, b's, and c's. Each word represents a symmetry moving a given triangle via a sequence of reflections to a copy of that triangle elsewhere in the plane.

Cutting and Pasting Triangles

A triangle with sides 4, 12, 12 and a triangle with sides 6, 9, 13 both have perimeter 28 and area $4\sqrt{35}$ (by Heron's formula). Is there a way to cut the first triangle into a finite number of pieces that can be reassembled to form the second triangle, so that the perimeter of the first triangle becomes the perimeter of the second triangle?

Solution

Surprisingly, given *any* two triangles with the same perimeter and area, there is a way to cut the first triangle into a finite number of pieces that can be reassembled to form the second triangle, so that the perimeter of the first becomes the perimeter of the second. Observe that the radius r of the inscribed circle of a triangle satisfies $rs = K$, where s is the semiperimeter of the triangle and K is its area. (See Toolkit.) This is obvious from the diagram below. Hence $r = K/s$, so that the two given triangles have the same inradius.

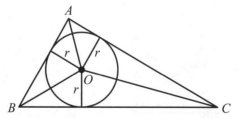

Our method of cutting the first triangle to form the second triangle is also inherent in the above diagram. Cut the triangle $\triangle ABC$ into triangles $\triangle AOB$, $\triangle BOC$, and $\triangle COA$, where O is the incenter. Lay the three triangular pieces along a line, as in the picture below.

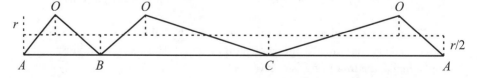

2.2 Geometry

In the picture, both endpoints of the line are labeled A to indicate how the triangles are cut and laid along the line. Also, the apexes of the triangles are all labeled O, since they all come from the incenter.

Cut each of the three triangles into three pieces and arrange the pieces to make a rectangle of the same area. This is easy to do by cutting along the dotted lines in the diagram. Note that the altitude of each triangle (from the line) is interior to the triangle. The base of each triangle becomes the base of the corresponding rectangle. If we perform this procedure for both of the given triangles, then we obtain two rectangles with base $2s$ and height $r/2$. We re-cut the rectangle from the first triangle to match the cuts for the second triangle, and then form the three subtriangles for the second triangle. This solves the problem.

Bonus: A Family of Triangle Pairs

Let's find an infinite family of non-similar pairs of triangles with integer side lengths such that the two triangles in each pair have the same perimeter and the same area. Suppose that the two triangles in a pair have sides a, b, c and a', b', c'. Since the triangles have the same perimeter, $2s = a + b + c = a' + b' + c'$. Because they have the same area, we have, by Heron's formula,

$$\sqrt{s(s-a)(s-b)(s-c)} = \sqrt{s(s-a')(s-b')(s-c')},$$

and hence,

$$(s-a)(s-b)(s-c) = (s-a')(s-b')(s-c').$$

As $(s-a) + (s-b) + (s-c) = 3s - 2s = s = (s-a') + (s-b') + (s-c')$, we must find triples of positive integers with the same sum and the same product. Set

$$s - a = p \qquad\qquad s - a' = 1$$
$$s - b = p \qquad\qquad s - b' = pq$$
$$s - c = q^2 \qquad\qquad s - c' = pq,$$

where p and q are positive integers greater than 1. The equal sum condition means that

$$2p + q^2 = 1 + 2pq,$$

so that

$$q = 2p - 1.$$

Hence $s = 4p^2 - 2p + 1$, and we have the pairs $\{4p^2 - 3p + 1, 4p^2 - 3p + 1, 2p\}$ and $\{4p^2 - 2p, 2p^2 - p + 1, 2p^2 - p + 1\}$, with common perimeter $8p^2 - 4p + 2$ and common area $p(2p-1)\sqrt{4p^2 - 2p + 1}$. For example, with $p = 3$, we have the pairs $\{30, 16, 16\}$ and $\{28, 28, 6\}$, both with perimeter 62 and area $15\sqrt{31}$.

Our problem is similar to van Schooten's Problem:[3] Given an isosceles triangle, find another isosceles triangle with the same perimeter and the same area. Isaac Newton (1642–1727) gave a solution based on the intersection of a hyperbola and a parabola.

---

[3] The problem is named after Franz van Schooten (1615–1660).

Cookie Cutting

Suppose that we have an equilateral triangle made out of cardboard. Can we cut the triangle into a finite number of polygonal pieces that can be reassembled to form a square (of the same area)?

Solution

Yes, it can be done, and in fact a solution due to Henry Ernest Dudeney (1857–1930) uses just four pieces. We can "see" the dissection via a simultaneous tessellation of the plane by squares and by equilateral triangles.

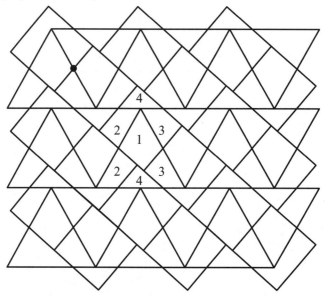

The labeling of a particular equilateral triangle by 1, 2, 3, 4 shows how the triangle can be dissected into four pieces that can be reassembled to form the square.

In describing the tessellation precisely, we work with parallelograms such as those centered around the large dot in the picture. The parallelogram formed from two adjacent squares is inclined at a slope of $-\sqrt{3}/2$ from the parallelogram formed from two adjacent equilateral triangles.

Bonus: The Wallace–Bolyai–Gerwien Theorem

If A and B are any plane polygonal regions of the same area, then A can be cut into a finite number of pieces that can be reassembled to form B. This is known as the Wallace–Bolyai–Gerwien theorem. We will sketch a proof.

As in the Solution to "Cutting and Pasting Triangles," any triangle can be cut into three pieces that can be reassembled to make a rectangle of the same area. Cut the triangle along an internal altitude and along the perpendicular bisector of that altitude. It's easy to arrange the pieces to make a rectangle.

The next step in proving the Wallace–Bolyai–Gerwien theorem is showing that any rectangle can be transformed into a square of the same area. The picture below shows how this can be accomplished.

2.2 Geometry

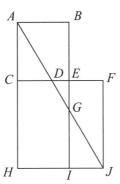

The rectangle $ABIH$ can be transformed into a square $CFJH$ of the same area. We leave it as an exercise to show that the triangle ACD is congruent to the triangle GIJ and the triangle ABG is congruent to the triangle DFJ. Now the transformation is obvious. If the rectangle is "too skinny," so that the point D is not within the rectangle, then the construction needs tweaking. Cut the rectangle in half (the short way) and stack the two pieces together to make a "fatter" rectangle. Repeat this tweaking if necessary.

The above argument can be modified to transform any rectangle into a rectangle with a prescribed base.

Now we can "put the pieces together" and complete the proof of the Wallace–Bolyai–Gerwien theorem. Cut the polygonal region A into triangles. Dissect the triangles into squares. Cut the squares into pieces and rearrange them into rectangles all with the same base. Put these rectangles together to form one rectangle. Transform that rectangle into a square. Now, do the same process for B. Since both A and B can be cut into pieces that can be reassembled to make squares of the same area, we can cut A into pieces and reassemble the pieces to make B.

By the way, the corresponding dissection problem in three dimensions cannot always be achieved. In particular, there exist two tetrahedra of the same volume such that one tetrahedron cannot be cut into a finite number of pieces that can be reassembled to make the other tetrahedron. This answers in the negative a famous problem of David Hilbert (Hilbert's Third Problem). For more on Hilbert's problems, see p. 161.

Revolving Credit

A certain credit card has the shape of a rectangle of dimensions 3 units × 4 units. If you rotate the card about one of its diagonals, what is the volume of the resulting solid of revolution?

Here is a picture of the solid.

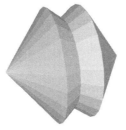

Solution

We solve the general problem for a rectangular card of width w and length l, where $w \leq l$. Consider the diagram below.

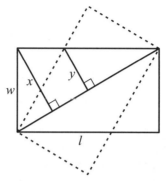

Using the formula for the volume of a cone (1/3 base × height), we have

$$\text{volume} = 2\left(\frac{1}{3}\pi x^2 \sqrt{w^2 + l^2} - \frac{1}{3}\pi y^2 \frac{\sqrt{w^2 + l^2}}{2}\right)$$

$$= \frac{\pi}{3}\sqrt{w^2 + l^2}(2x^2 - y^2).$$

By similar triangles, $x/w = l/\sqrt{w^2 + l^2}$, and so $x^2 = w^2 l^2/(w^2 + l^2)$. Likewise, $y/(\sqrt{w^2 + l^2}/2) = w/l$, and so $y^2 = w^2(w^2+l^2)/(4l^2)$. Substituting in the expressions for x^2 and y^2, and simplifying, we obtain

$$\text{volume} = \frac{\pi w^2 (7l^4 - 2l^2 w^2 - w^4)}{12 l^2 \sqrt{w^2 + l^2}}.$$

For $w = 3$ and $l = 4$, we calculate that the volume of revolution is $4269\pi/320$ cubic units.

Bonus: The Case of the Rotating Parallelogram

What is the volume obtained by rotating a parallelogram about one of its diagonals? Assume that the parallelogram has side lengths a and b, with $a \leq b$, and diagonal lengths d_1 and d_2. Of course, the four parameters are interdependent. Using the law of cosines (see Toolkit), they can be shown to satisfy

$$d_1^2 + d_2^2 = 2a^2 + 2b^2.$$

Let's assume that $b - a \leq d < a + b$, where d is the diagonal about which the parallelogram is rotated. Consider the diagram below.

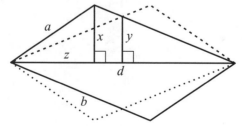

2.2 Geometry

We find that $x^2 + z^2 = a^2$ and $x^2 + (d-z)^2 = b^2$. Taking the difference of these equations, solving for z, and plugging the expression for z into the first equation yields

$$x^2 = a^2 - \left(\frac{a^2 - b^2 + d^2}{2d}\right)^2.$$

By similar triangles, $y/(d/2) = x/(d-z)$. Hence

$$y^2 = \frac{d^2 x^2}{4(b^2 - x^2)}.$$

The volume is given by

$$\frac{\pi d}{3}(2x^2 - y^2).$$

Upon substitution, the volume is

$$v(a, b, d) = \frac{\pi d}{3}\left(a^2 - \frac{(a^2 - b^2 + d^2)^2}{4d^2}\right)\left(2 - \frac{d^4}{(a^2 - b^2 - d^2)^2}\right).$$

For example,

$$v(3, 4, 5) = 4269\pi/320$$

(the same result as in the Solution). Note that the volume formula also holds in the case where the parallelogram leans over the diagonal and the resulting solid is concave.

What is the maximum volume, given a and b? When I posed this problem, I thought that the maximum volume would occur when the two sides of the parallelogram are perpendicular, or perhaps when the short side is perpendicular to a diagonal. But I was due for a surprise, as neither is true. In fact, the optimal value of d^2 is a solution to a quintic (degree 5) polynomial! I didn't expect to see a quintic come up in this problem.

The derivative of the volume function with respect to d is a rational function of a, b, and d. We can write this rational function (divided by π) as a fraction whose denominator is positive and whose numerator is

$$h(a, b, d) = -2(a^2 - b^2)^5 + 2(a^2 - 5b^2)(a^2 - b^2)^3 d^2$$
$$+ 3(a^2 - b^2)^2(3a^2 + 5b^2)d^4 - (a^2 - b^2)(19a^2 - 15b^2)d^6$$
$$+ (13a^2 - 9b^2)d^8 - 3d^{10}.$$

Since the denominator is positive, the derivative changes sign exactly when h does. If we set this polynomial (a quintic in d^2) equal to 0, we will find candidates for the optimal value of d.

Right away, we can verify that the maximum volume *does not* occur when the sides are perpendicular, or when one side is perpendicular to a diagonal. Using a computer or calculator, we find that there is one positive real solution to $h(3, 4, d) = 0$, namely, $d \approx 3.70407$. Now we compare the volumes based on d having this value, d being the side length of a diagonal of a rectangle, and d being the side length of a diagonal perpendicular to the shorter side:

$$\text{volume}(3, 4, 3.70407) \approx 49.5291$$

$$\text{volume}(3, 4, 5) \approx 41.9108$$

$$\text{volume}(3, 4, \sqrt{7}) \approx 43.6373.$$

We see that the optimal diagonal is not the diagonal of a rectangle nor is it perpendicular to a side of the parallelogram.

We will now prove that the volume function has a unique local maximum. We do this by showing that the derivative is positive up to a certain point, and thereafter the curve is concave down.

By scaling, we may as well assume that $a = 1$.

Using Descartes' rule of signs (see Toolkit) we find that h has exactly one positive real root if b^2 is in the ranges

$$1 \leq b^2 \leq 19/15 \quad \text{or} \quad 13/9 \leq b^2.$$

So let's assume that b^2 falls in the range

$$19/15 < b^2 < 13/9.$$

Now we look at the second derivative of the volume function. Upon clearing the denominator (which is positive) and factoring out π, we obtain

$$-2(1-b^2)^6 + 8(1-b^2)^5 d^2$$
$$- 15(1-b^2)^4 d^4 + 20(1-3b^2)(1-b^2)^2 d^6$$
$$- 4(1-b^2)(5-3b^2)d^8 + 12(1-b^2)d^{10} - 3d^{12}.$$

Since $1 \leq b$, the coefficients of d^0, d^2, d^4, d^6, and d^{12} are negative. Moreover, the sum of the degree 8 and 10 terms is negative under a certain condition. The sum of the d^8 and d^{10} terms is

$$-4(5-3b^2)(1-b^2)d^8 + 12d^2(1-b^2)d^8 = (-20 + 12b^2 + 12d^2)(1-b^2)d^8.$$

Hence, the volume function is concave down when

$$-20 + 12b^2 + 12d^2 > 0.$$

Since we are assuming that $b^2 > 19/15$, this will hold if $d^2 > 2/5$. So let's now assume that

$$d^2 < 2/5.$$

We will prove that the volume function is increasing on this interval ($b-1 \leq d \leq \sqrt{2/5}$). This is equivalent to showing that the function $-h$ is positive. With our assumptions, the coefficients of d^6 and d^{10} are positive, while the coefficients of d^0, d^2, d^4, and d^8 are negative.

Let's combine the d^4, d^6 and d^{10} terms and divide by d^4. If we replace d with $\sqrt{2/5}$ we get the following upper bound on this expression:

$$-\frac{23}{25} - \frac{53}{5}b^2 + 27b^4 - 15b^6.$$

We claim that this expression is negative when $19/15 < b^2 < 13/9$. Take the derivative:

$$-\frac{106}{5}b + 108b^3 - 90b^5.$$

Again, look at the last two terms. The sum is negative if and only if $108/90 < b^2$. We have $108/90 < 19/15 < b^2$, so the expression is decreasing. Since the value at $b^2 = 19/15$ is negative, our claim is proved.

Given a triangle with side lengths a and b, with $a \leq b$, what should be the third side length c, so that if the triangle is rotated about the third side, the volume of the resulting body of revolution is maximized? Of course, c will satisfy $b - a \leq c \leq b + a$.

2.3 Calculus

The Harmonic Series

Show that the *harmonic series*

$$1 + \frac{1}{2} + \frac{1}{3} + \frac{1}{4} + \frac{1}{5} + \frac{1}{6} + \frac{1}{7} + \frac{1}{8} + \cdots$$

has an infinite sum.

Solution

Consider the sum of the first 2^k terms (where $k \geq 1$):

$$1 + \frac{1}{2} + \frac{1}{3} + \frac{1}{4} + \frac{1}{5} + \frac{1}{6} + \frac{1}{7} + \frac{1}{8} + \cdots + \frac{1}{2^k}.$$

This sum is greater than

$$1 + \frac{1}{2} + \left(\frac{1}{4} + \frac{1}{4}\right) + \left(\frac{1}{8} + \frac{1}{8} + \frac{1}{8} + \frac{1}{8}\right) + \cdots + \left(\frac{1}{2^k} + \cdots + \frac{1}{2^k}\right).$$

Since each series in parentheses adds up to $1/2$, this latter sum is

$$1 + \frac{k}{2},$$

which tends to infinity with k. It follows that the harmonic series has an infinite sum.[4]

Bonus: Estimating the Harmonic Sum

Let

$$H_n = 1 + \frac{1}{2} + \frac{1}{3} + \cdots + \frac{1}{n}.$$

The above argument shows that

$$H_{2^k} > 1 + \frac{k}{2}.$$

Hence, if $n = 2^k$, then

$$H_n > 1 + \frac{\log_2 n}{2}.$$

---

[4] This argument was invented by Nicole Oresme (c. 1323–1382).

Using the method of comparison of areas, we can obtain for H_n both an upper bound and a better lower bound. Since the function $1/x$ is decreasing for $x > 0$, we have

$$\int_1^{n+1} \frac{dx}{x} < H_n < 1 + \int_1^n \frac{dx}{x}.$$

Evaluating the integrals, we obtain

$$\ln(n+1) < H_n < 1 + \ln n.$$

Therefore, $H_n \sim \ln n$ (i.e., H_n is asymptotic to $\ln n$). This means that $H_n/\ln n \to 1$ as $n \to \infty$.

A Quick Integral

Evaluate the definite integral

$$\int_0^1 \ln x \, dx.$$

Solution

By symmetry (see figure),

$$\int_0^1 \ln x \, dx = -\int_{-\infty}^0 e^x \, dx.$$

Knowing that e^x is its own antiderivative, we can complete the calculation:

$$\int_0^1 \ln x \, dx = -\int_{-\infty}^0 e^x \, dx = -\left(e^x\big]_{-\infty}^0\right) = -\left(e^0 - e^{-\infty}\right) = -(1-0) = -1.$$

2.3 Calculus

Bonus: Another Quick Integral

We can also easily evaluate

$$\int_0^1 \arcsin x \, dx.$$

By symmetry, this integral equals the area of a rectangle minus the area under a sine curve:

$$1 \cdot \frac{\pi}{2} - \int_0^{\pi/2} \sin x \, dx = \frac{\pi}{2} - (-\cos x)]_0^{\pi/2} = \frac{\pi}{2} + \cos(\pi/2) - \cos 0 = \frac{\pi}{2} - 1.$$

Euler's Sum

Evaluate the infinite sum

$$1 + \frac{1}{4} + \frac{1}{9} + \frac{1}{16} + \frac{1}{25} + \frac{1}{36} + \frac{1}{49} + \cdots.$$

Note. The numerical value was determined by Leonhard Euler[5] (1707–1783).

Solution

Euler's solution is a bit of magic, starting with the power series representation of $\sin x$:

$$\sin x = x - \frac{x^3}{3!} + \frac{x^5}{5!} - \frac{x^7}{7!} + \cdots.$$

Euler reasoned that since the zeros of the sine function are the integer multiples of π (i.e., $n\pi$, where n is an integer), $\sin x$ has the infinite product expansion

$$\sin x = x \left(1 - \frac{x}{\pi}\right)\left(1 + \frac{x}{\pi}\right)\left(1 - \frac{x}{2\pi}\right)\left(1 + \frac{x}{2\pi}\right)\left(1 - \frac{x}{3\pi}\right)\left(1 + \frac{x}{3\pi}\right) \cdots.$$

To see this, simply plug in $x = n\pi$ and note that the right side is 0 (for any integer n). The magic of the proof is that although this reasoning is valid for a polynomial, $\sin x$ is not a polynomial. Euler's method is a leap of faith.

We equate our two representations:

$$x - \frac{x^3}{3!} + \frac{x^5}{5!} - \frac{x^7}{7!} + \cdots = x\left(1 - \frac{x^2}{\pi^2}\right)\left(1 - \frac{x^2}{4\pi^2}\right)\left(1 - \frac{x^2}{9\pi^2}\right) \cdots.$$

Now consider the coefficient of x^3. On the left side, it is $-1/3! = -1/6$. On the right side, it is

$$-\frac{1}{\pi^2}\left(1 + \frac{1}{4} + \frac{1}{9} + \cdots\right).$$

Since these coefficients must be the same, we have

$$1 + \frac{1}{4} + \frac{1}{9} + \cdots = \frac{\pi^2}{6}.$$

This is a totally unexpected answer!

---

[5] Leonhard Euler was one of the most prolific mathematicians of all time. He made foundational contributions in analysis, number theory, graph theory, and mathematical notation.

Bonus: Probability that Two Numbers are Coprime

Suppose that two positive integers from 1 to n are chosen at random. As n tends to infinity, what is the probability that the integers are coprime (have no factor in common)? Two numbers have a factor in common if and only if they have a prime factor in common. The probability that two integers are not both divisible by 2 is, in the limit, $1 - 1/2^2$. The probability that they are not both divisible by 3 is $1 - 1/3^2$. In general, the probability that two integers are not both divisible by a prime p is $1 - 1/p^2$. Hence, the probability that the two integers are coprime is, in the limit,

$$\prod_p \left(1 - \frac{1}{p^2}\right),$$

where the product is taken over all prime numbers p. Using the formula for the sum of a geometric series (see p. 16), we can write this expression as

$$\prod_p \frac{1}{1 + \frac{1}{p^2} + \frac{1}{p^4} + \cdots} = \frac{1}{\prod_p \left(1 + \frac{1}{p^2} + \frac{1}{p^4} + \cdots\right)}.$$

Since the reciprocal of each square integer is accounted for exactly once in the product, this expression is equal to

$$\frac{1}{1 + \frac{1}{4} + \frac{1}{9} + \cdots},$$

the reciprocal of Euler's sum. Therefore, the probability that two positive integers chosen at random are coprime is $6/\pi^2 \approx 0.608$. Did you expect to see π in the answer to this problem?

Strips of Carpeting

Suppose that the unit disk is covered with n infinitely long rectangular strips of widths w_i, where $1 \leq i \leq n$. Prove that the sum of the w_i is at least 2.

Solution

We assume that the strips don't extend "width-wise" outside the disk (because any that did would only contribute to a larger sum of the widths). Let S be the unit sphere with the same center as the given circle and project each strip onto S. (A bolt from the blue!)

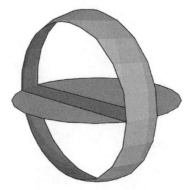

2.3 Calculus

We claim that the surface area of the projection of the ith strip is $2\pi w_i$. (Isn't it interesting that the area depends only on the width of the strip and not where it is placed?) To see this, we compute the surface area of the projection of a strip as an area of revolution. Choose the coordinate axes so that the ith strip is perpendicular to the x-axis. Suppose that the ith strip is bounded by the lines $x = a$ and $x = b$. Then the surface area is given by

$$\int_a^b 2\pi y \sqrt{1 + \frac{x^2}{y^2}}\, dx = 2\pi(b-a) = 2\pi w_i.$$

Since the disk is covered by the strips, the surface of the sphere is also covered. It follows that $\sum_{i=1}^n 2\pi w_i \geq 4\pi$ and $\sum_{i=1}^n w_i \geq 2$.

Bonus: Four Spherical Triangles

The above analysis shows that, for example, the surface areas of the following four regions on a spherical globe are identical: (1) The region between 30° North Latitude and the North Pole; (2) The region between 30° North Latitude and the equator; (3) The region between the equator and 30° South Latitude; (4) The region between 30° South Latitude and the South Pole. If the globe has surface area 4π, then each of these four regions has surface area π.

What if we start with a *spherical triangle* of surface area π (on a sphere of surface area 4π)? Is it true that four congruent copies of the spherical triangle can be arranged to cover the sphere? This turns out to be the case. Since the angle sum of the given spherical triangle is 2π, we can place, at any point on the sphere, three copies of the triangle meeting at a vertex. The three triangles define a complementary fourth region on the sphere that is clearly congruent to the same triangle (see [4]).

Is it 0?

Given

$$f(x) = \frac{607}{607x + 1} + \frac{692}{692x + 1} + \frac{-701}{-701x + 1},$$

determine, without using a computing device, whether or not

$$f^{(12)}(0) = 0,$$

where $f^{(12)}$ denotes the twelfth derivative of f.

Solution

Via the power rule and the chain rule for derivatives, we obtain

$$f^{(12)}(x) = 12! \left[\left(\frac{607}{607x+1}\right)^{13} + \left(\frac{692}{692x+1}\right)^{13} + \left(\frac{-701}{-701x+1}\right)^{13} \right].$$

Hence

$$f^{(12)}(0) = 12!(607^{13} + 692^{13} - 701^{13}).$$

Is it possible that this quantity is equal to 0? No, by Fermat's Last Theorem (proved by Andrew Wiles)! Remember that Fermat's Last Theorem says that the equation

$$a^n + b^n = c^n$$

is not solvable in positive integers a, b, c, if $n > 2$.

Bonus: Fermat's Last Theorem

Fermat's Last Theorem (FLT) was the most famous unsolved mathematics problem when Andrew Wiles solved it in 1995. The problem originated in 1637 when Pierre de Fermat (1601–1665) wrote in the margin of his copy of Diophantus' *Arithmetic*,

> Cubum autem in duos cubos, aut quadratoquadratum in duos quadratoquadratos & generaliter nullam in infinitum ultra quadratum potestatem in duos eiusdem nominis fas est diuidere cuius rei demonstrationem mirabilem sane detexi. Hanc marginis exiguitas non caperet.

> [It is impossible to separate a cube into two cubes, or a fourth power into two fourth powers, or in general, any power higher than the second into two like powers. I have discovered a truly marvelous proof of this, which this margin is too narrow to contain.]

Wiles' proof is based on proving the far-reaching Taniyama–Shimura conjecture concerning elliptic curves.[6]

Pi is Pi

The constant π appears in the formula for the circumference of a circle ($C = 2\pi r$) and in the formula for the area of a circle ($A = \pi r^2$), where r is the radius of the circle. Suppose that π is *defined* by the circumference formula. Prove the area formula *without using circular reasoning*.

Solution

The catch is the *without using circular reasoning* clause. For example, there is a flaw in the following argument. Since the circle is given by the equation $x^2 + y^2 = r^2$, the area of the circle is given by the integral

$$2 \int_{-r}^{r} \sqrt{r^2 - x^2} \, dx.$$

---

[6] I highly recommend Simon Singh's documentary film *The Proof*, about the proof of FLT.

2.3 Calculus

Making the change of variables $x = r \sin \theta$, this integral becomes

$$2 \int_{-\pi/2}^{\pi/2} r^2 \cos^2 \theta \, d\theta = \int_{-\pi/2}^{\pi/2} r^2 (1 + \cos 2\theta) \, d\theta$$

$$= \pi r^2 + r^2 \int_{-\pi/2}^{\pi/2} \cos 2\theta \, d\theta$$

$$= \pi r^2 + r^2 \left(\frac{1}{2} \sin 2\theta \right) \bigg|_{-\pi/2}^{\pi/2}$$

$$= \pi r^2.$$

The flaw in the reasoning is that the fact that the derivative of $\sin \theta$ is $\cos \theta$ (used in two places above) depends on the formula for the area of a circle.

Here is an argument that avoids circular reasoning. The circumference formula, written as an integral for arc length, is

$$2\pi r = 2 \int_{-r}^{r} \sqrt{1 + (y')^2} \, dx = 2 \int_{-r}^{r} \frac{r}{\sqrt{r^2 - x^2}} \, dx.$$

Hence, using the fact that $x \, dx = -y \, dy$, we obtain

$$\pi r^2 = \int_{-r}^{r} \frac{r^2}{\sqrt{r^2 - x^2}} \, dx$$

$$= 2 \int_{0}^{r} \frac{r^2 - x^2 + x^2}{\sqrt{r^2 - x^2}} \, dx$$

$$= 2 \int_{0}^{r} \sqrt{r^2 - x^2} \, dx + 2 \int_{0}^{r} \frac{x^2}{\sqrt{r^2 - x^2}} \, dx$$

$$= 2 \int_{0}^{r} \sqrt{r^2 - x^2} \, dx + 2 \int_{r}^{0} \frac{r^2 - y^2}{y} \left(-\frac{y}{x} \right) dy$$

$$= 2 \int_{0}^{r} \sqrt{r^2 - x^2} \, dx + 2 \int_{0}^{r} \sqrt{r^2 - y^2} \, dy$$

$$= 4 \int_{0}^{r} \sqrt{r^2 - x^2} \, dx$$

$$= A.$$

This establishes the area formula.

Bonus: Irrationality of π

We will give a concise proof by contradiction, due to Ivan Niven (1915–1999), that π is irrational.

Suppose that π is a rational number, say, $\pi = p/q$, where p and q are positive integers. Define the polynomial

$$f(x) = \frac{x^n (p - qx)^n}{n!},$$

where n is to be determined. We note that $f(x) = f(\pi - x)$, for all x. It follows that $f^{(i)}(0) = \pm f^{(i)}(\pi)$, for all $i \geq 0$. We claim that $f^{(i)}(0)$ (and hence also $f^{(i)}(\pi)$) is an integer for each $i \geq 0$. For a typical term of $f(x)$ looks like

$$\frac{\alpha}{n!} x^{n+k},$$

where $0 \leq k \leq n$ and α is an integer. Upon differentiation i times and evaluation at 0, this quantity is 0 if $i \neq n + k$ and $\alpha \frac{(n+k)!}{n!} = \alpha k! \binom{n+k}{n}$ if $i = n + k$.

Define

$$F(x) = f(x) - f^{(2)}(x) + f^{(4)}(x) - \cdots + (-1)^n f^{(2n)}(x).$$

Observe that

$$\frac{d}{dx}[F'(x) \sin x - F(x) \cos x] = [F''(x) + F(x)] \sin x$$

$$= [f(x) + (-1)^{n+2} f^{(2n+2)}(x)] \sin x$$

$$= f(x) \sin x$$

(since f has degree $2n$).[7] Now we compute

$$\int_0^\pi f(x) \sin x \, dx = [F''(x) \sin x - F(x) \cos x]\Big|_0^\pi = F(\pi) + F(0).$$

Since all $f^{(i)}(0)$ and $f^{(i)}(\pi)$ are integers, $F(\pi) + F(0)$ is an integer. However, for $0 < x < \pi$, we have

$$0 < f(x) \sin x < \frac{(\pi p)^n}{n!}.$$

In the upper bound, the numerator is an exponential function of n with a fixed base; this is dominated by the $n!$ function in the denominator, as n tends to infinity. Hence, the integral is positive but arbitrarily small for sufficiently large n. But this is impossible since it is an integer. So our assumption that π is a rational number is false. Therefore, π is irrational.

2.4 Probability

How Many Birthdays?

There are 100 people in a room. Each person states his/her birth date (month and day, excluding February 29). Let N be the number of different birth dates. What is the expected value of N? (Assume that the 365 birth dates are equally likely.)

---

[7] The definition of $F(x)$ is motivated by our need for the relation $F'' + F = f$. Working backwards from this relation, we obtain $(I + D^2)F = f$, where I is the identity operator and D is the derivative operator. Hence $F = \left(\frac{1}{I+D^2}\right) f = f - f^{(2)} + f^{(4)} - \cdots$.

2.4 Probability

Solution

For $1 \leq i \leq 365$, let $X_i = 1$ if date i is among the stated birth dates and $X_i = 0$ otherwise. Then the expected number of different birth dates among the 100 people is

$$E(X_1 + X_2 + X_3 + \cdots + X_{365}) = E(X_1) + E(X_2) + E(X_3) + \cdots + E(X_{365}),$$

because of linearity of expectation (independence of events is not necessary). Since all the dates are interchangeable, the sum on the right is equal to $365 \cdot E(X_1)$. The value of $E(X_1)$ is the probability that January 1 is among the 100 dates, which is the complementary probability of the event that January 1 is *not* one of the 100 dates. Hence $E(X_1) = 1 - (364/365)^{100}$. Therefore, the expected number of different birth dates among 100 people is

$$365(1 - (364/365)^{100}) \approx 88.$$

Bonus: Average Number of Fixed Points of a Permutation

In a random permutation of n objects, how many objects are fixed (not moved by the permutation)? This problem is also solved simply by using expectation. For $1 \leq i \leq n$, let $X_i = 1$ if object i is fixed by the permutation and $X_i = 0$ otherwise. Then the expected number of fixed objects is

$$E(X_1 + X_2 + X_3 + \cdots + X_n) = E(X_1) + E(X_2) + E(X_3) + \cdots + E(X_n)$$
$$= \frac{1}{n} + \frac{1}{n} + \frac{1}{n} + \cdots + \frac{1}{n} \quad (n \text{ terms})$$
$$= 1.$$

On average, one object is fixed, regardless of n.

The Average Number of Spots

A fair die is rolled until a 6 occurs. What is the average total number of spots seen on the die before the 6 occurs? For example, if the rolls are 4, 2, 5, 2, 6, then the number is $4 + 2 + 5 + 2 = 13$.

Solution

The average number of spots is 15. Suppose that the random variable X is the number of spots seen on the die before the 6 occurs. Then the expected value of X, denoted $E(X)$, satisfies the equation

$$E(X) = \frac{E(X) + 1}{6} + \frac{E(X) + 2}{6} + \frac{E(X) + 3}{6} + \frac{E(X) + 4}{6} + \frac{E(X) + 5}{6}.$$

The reason is that the first roll is 1, 2, 3, 4, 5, or 6, each possibility occurring with probability $1/6$, and these outcomes add 1, 2, 3, 4, 5, or 0, respectively, to the total number of spots.

Solving the equation for $E(X)$ yields $E(X) = 15$.

Bonus: Average Waiting Time

How are we to interpret the answer 15 in our problem? The answer is related to a basic probability problem called *waiting time*. Suppose that an event happens with probability p. For example, in our problem, a die comes up 6 with probability $1/6$. How long, on average, do we have to wait for the event to happen, given repeated, independent trials? For example, if the die comes up 4, 2, 5, 2, 6, then the number of trials is 5. Let the expected number of trials be E. Then E satisfies the equation

$$E = p \cdot 1 + (1-p)(E+1).$$

The reasoning is similar to that in the Solution. Solving for E, we obtain $E = 1/p$. In the case of the die, this means that $E = 6$. So the average number of rolls *before* the 6 occurs is 5. Since the number of spots on one roll of the die when it does not come up 6 is on average $(1+2+3+4+5)/5$, it follows that the average total number of spots before a 6 occurs is $5 \cdot (1+2+3+4+5)/5 = 1+2+3+4+5 = 15$.

Balls Left in an Urn

An urn contains n white balls and n black balls. A ball is chosen at random and removed. This process is repeated until the urn contains only balls of one color. What is the expected number of balls remaining in the urn?

Solution

We first pursue a mundane method using binomial coefficients. Suppose that there are k black balls left in the urn, where $1 \leq k \leq n$. (The case of k white balls is similar.) The situation of k black balls left in the urn arises when, among the first $2n-k-1$ balls chosen, there are $n-1$ white balls and $n-k$ black balls, and the next (i.e., $(2n-k)$th) ball chosen is white. The probability that the $(2n-k)$th ball is white is $1/(k+1)$. Thus, the expected number of balls left in the urn is

$$E = 2 \sum_{k=1}^{n} k \cdot \frac{\binom{n}{n-1}\binom{n}{n-k}}{\binom{2n}{2n-k-1}} \cdot \frac{1}{k+1}.$$

For $n = 1, 2, 10$, and 100, we find that $E = 1, 4/3, 20/11$, and $200/101$, respectively. So we guess that the general formula is $E = 2n/(n+1)$. Are you surprised that the answer tends to 2 as n goes to infinity?

An aha! proof relies on two key realizations. First, the term $2n$ in the solution tells us that we should figure out the probability that a given ball is left in the urn and multiply this by $2n$. Now, we need to show that the probability that a given ball is left in the urn is $1/(n+1)$. So what is the probability that a given white ball is left in the urn? The second aha! realization is to imagine the experiment continuing until *all* the balls are selected and withdrawn. Then the given white ball is left in the urn (in the original problem) if and only if it is selected after all the black balls in the extended problem. There are $n+1$ turns at which the white ball can be selected (relative to the black balls and ignoring all the other white balls): before the first black ball, between the first and second black balls, ..., after

2.4 Probability

the last black ball. Since these turns are equally likely, the probability that the white ball is left in the urn is $1/(n+1)$, and we are done.

As an exercise, generalize the problem so that the urn starts with b black balls and w white balls. Suppose that balls are withdrawn at random until only balls of one color remain. Show that the expected number of balls remaining is $b/(w+1) + w/(b+1)$.

Bonus: A Two-Urn Problem

There are two urns. Urn A contains 5 white balls. Urn B contains 4 white balls and 1 black ball. An urn is selected at random and a ball in that urn is selected at random and removed. This procedure is repeated until one of the urns is empty. What is the probability that the black ball has not been selected?

Actually, the problem isn't difficult to solve. The difficulty is in interpreting the answer.

The favorable outcomes end with a ball chosen from Urn A. The number of balls selected from Urn B can be any k where $0 \leq k \leq 4$, meaning that $5+k$ balls are selected altogether. In making $5 + k$ selections of the urns, the probability of selecting Urn A 5 times and Urn B k times, with Urn A selected last, is $(1/2)^{4+k} \binom{4+k}{k}(1/2)$. Given that k balls are selected from Urn B, the probability that the black ball is not selected is $(5-k)/5$. Therefore, the probability of a favorable outcome is

$$\sum_{k=0}^{4} \binom{4+k}{k} \left(\frac{1}{2}\right)^{5+k} \left(\frac{5-k}{5}\right) = \frac{63}{256}.$$

Scrutinizing the answer, we notice that the denominator is a power of 2, i.e., 2^8. If we change it to 2^{10} (so that the exponent is equal to the total number of balls), then the numerator becomes 252, which is $\binom{10}{5}$. So the answer may be written

$$\frac{\binom{10}{5}}{2^{10}}.$$

The presence of a binomial coefficient and a power of 2 makes us ask, what is being counted?

The problem may be generalized so that Urn A contains n white balls and Urn B contains $n-1$ white balls and 1 black ball. The probability that the black ball is not selected is given by

$$\sum_{k=0}^{n} \binom{n-1+k}{k} \left(\frac{1}{2}\right)^{n+k} \left(\frac{n-k}{n}\right).$$

(Notice that using an upper summation limit of $k = n$ instead of $k = n-1$ only introduces a 0 summand.) The sum may be written as a telescoping series:

$$\sum_{k=1}^{n} \left[\binom{n+k}{k} \left(\frac{1}{2}\right)^{n+k} - \binom{n-1+k}{k-1} \left(\frac{1}{2}\right)^{n-1+k} \right] + \left(\frac{1}{2}\right)^n = \frac{\binom{2n}{n}}{2^{2n}}.$$

So in general we ask, what is being counted?

Here are three intriguing variations of the above problem.

Tie Series: The first variation is a non-probability version. Suppose that teams A and B play a series of six games with each game won by one of the teams. We can represent the outcomes of the games by a string of A's and B's, showing who wins each game. Three examples of strings are

$$AAAAAB$$
$$AABBAB$$
$$BBABBB.$$

The string AAAAAB, for instance, signifies that team A wins the first five games and team B wins the last game. Since there are two possible outcomes for each of the six games, there are $2^6 = 64$ possible strings.

Let's suppose that the winner of the series is decided as soon as one of the teams wins three games. Any games after that are considered exhibition games. We put an asterisk (*) in the string at the point where the series is decided. Thus, the three example strings above become

$$AAA*AAB$$
$$AABBA*B$$
$$BBAB*BB.$$

Let's perform this procedure for all of the 64 series.

AAA*AAA	ABAA*AA	BAAA*AA	BBAAA*A
AAA*AAB	ABAA*AB	BAAA*AB	BBAAA*B
AAA*ABA	ABAA*BA	BAAA*BA	BBAAB*A
AAA*ABB	ABAA*BB	BAAA*BB	BBAAB*B
AAA*BAA	ABABA*A	BAABA*A	BBAB*AA
AAA*BAB	ABABA*B	BAABA*B	BBAB*AB
AAA*BBA	ABABB*A	BAABB*A	BBAB*BA
AAA*BBB	ABABB*B	BAABB*B	BBAB*BB
AABA*AA	ABBAA*A	BABAA*A	BBB*AAA
AABA*AB	ABBAA*B	BABAA*B	BBB*AAB
AABA*BA	ABBAB*A	BABAB*A	BBB*AAA
AABA*BB	ABBAB*B	BABAB*B	BBB*ABB
AABBA*A	ABBB*AA	BABB*AA	BBB*BAA
AABBA*B	ABBB*AB	BABB*AB	BBB*BAB
AABBB*A	ABBB*BA	BABB*BA	BBB*BBA
AABBB*B	ABBB*BB	BABB*BB	BBB*BBB

We see that altogether there are 120 exhibition games, with 60 A wins and 60 B wins. What is surprising about this is that $120 = 6 \cdot 20 = 6 \cdot \binom{6}{3}$. Thus, the 60 A's and 60 B's can be arranged to form all the "tie series" of length 6, that is, strings with three A's and three B's. The twenty tie series of length 6 are

AAABBB	AABABB	BBABAA	BBBAAA
AABBAB	ABABAB	BABABA	BBAABA
AABBBA	ABBBAA	BBAAAB	BAAABB
ABBAAB	ABBABA	BAABBA	BAABBA.

2.4 Probability

In general, if we list all series of length $2n$, with the winning team the first to win n games, then the exhibition games can be arranged to produce all tie series of length $2n$. The tie series problem is equivalent to the urn problem since one can ask, in the series problem, what is the expected number of exhibition games per series? This is $2n$ times the probability that the black ball is not selected in the urn problem. I wonder whether there is a neat bijection proof of the tie series assertion.

Random City Walk: An object travels along the integer points of the plane, starting at the point $(0, 0)$. At each step, the object moves one unit to the right or one unit up (with equal probability). The object stops when it reaches the line $x = n$ or the line $y = n$. We will show that the expected length of the object's path is $2n - 2n\binom{2n}{n}2^{-2n}$.

The object's last stop before hitting a boundary line is a point of the form $(n-1, k)$ or $(k, n-1)$, for $0 \leq k \leq n-1$. The probability of visiting the point $(n-1, k)$ and then hitting the line $x = n$ is $\frac{1}{2}\binom{n+k-1}{n-1}$. By symmetry, the probability of visiting the point $(k, n-1)$ and then hitting the line $y = n$ is the same. Hence, the expected path length is given by

$$E = 2n - \sum_{k=0}^{n-1} (n-k) 2 \cdot \frac{1}{2} \binom{n+k-1}{n-1} \left(\frac{1}{2}\right)^{n+k-1}.$$

Thus, the expected number of steps to the corner (the point (n, n)) is

$$2n - E = \sum_{k=0}^{n} (n-k) \binom{n+k-1}{n-1} \left(\frac{1}{2}\right)^{n+k-1}.$$

This is a telescoping series, for the summand may be written

$$2n \binom{n+k}{n} \left(\frac{1}{2}\right)^{n+k} - 2n \binom{n+k-1}{n} \left(\frac{1}{2}\right)^{n+k-1}.$$

It follows that

$$E = 2n - 2n \binom{2n}{n} 2^{-2n}.$$

Is there an aha! proof of this formula?

Banach's Matchbox Problem:[8] A mathematician has two matchboxes, each containing n matches. He selects a matchbox at random and withdraws a match. He repeats this process until, attempting to withdraw a match, he finds that the matchbox he has selected is empty. In Banach's matchbox problem, one wants the probability distribution of the number of matches left in the nonempty box.

We can also ask, what is the expected number of matches in the nonempty box? Label the matches in the first box $1, 2, \ldots, n$, and those in the second box also $1, 2, \ldots, n$. The computation for the expected number of matches is basically the same as in the random walk problem. We can think of Banach's problem as concerning two matchboxes with $n + 1$ matches each. When the $(n + 1)$st match is taken, that means that the box has been emptied and the box has been selected again. We would like a simple proof that when the experiment is over, the probability that any particular match is still in its box is $\binom{2n}{n}2^{-2n}$; then the expected number of matches left is $2n$ times this quantity.

---

[8] The problem comes from Stefan Banach (1892–1945), the founder of functional analysis.

Random Points on a Circle

Given n random points on the circumference of a circle, what is the probability that they are all contained in a semicircular arc?

Solution

Let the points be $P_1, P_2, \ldots, P_n$. For $1 \leq i \leq n$, define E_i to be the event that the semicircular arc starting at P_i and going clockwise around the circle contains no P_j, for $j \neq i$. The events E_i are disjoint and $\Pr(E_i) = (1/2)^{n-1}$. Hence, the desired probability is

$$\Pr\left(\bigcup_{i=1}^{n} E_i\right) = \sum_{i=1}^{n} \Pr(E_i)$$
$$= \frac{n}{2^{n-1}}.$$

Bonus: Random Arcs on a Circle

Suppose that n arcs of length a are selected at random on a circle of circumference 1. With what probability is the circumference completely covered by the arcs?

Assuming that the arcs are given in counterclockwise order around the circle, it can be shown that the probability that there are "gaps" at the clockwise endpoints of i specified arcs is $(1 - ia)^{n-1}$ if $1 \leq i \leq \lfloor 1/a \rfloor$ and 0 otherwise. It follows by the principle of inclusion and exclusion (see Toolkit) that the probability that the circle is covered by the arcs is

$$\sum_{i=0}^{\lfloor 1/a \rfloor} (-1)^i \binom{n}{i} (1 - ia)^{n-1}.$$

This formula was found by W. L. Stevens.

The Gobbling Algorithm

Here is a procedure that I call the Gobbling Algorithm.

GOBBLING ALGORITHM. Start with a positive integer n. Choose an integer from 1 to n at random. Subtract this integer from n and replace n with the resulting integer. Repeat this process until you obtain 0.

Gobbling Algorithm

```
Let n be a positive integer.
While n > 0, do:
  Let k be a random positive integer between 1 and n.
  Output k.
  Let n ← n - k.
```

2.4 Probability

Let's do an example. Suppose that we start with $n = 10$. We choose an integer from 1 to 10 at random; let's say we choose 4. So we subtract, $10 - 4 = 6$. Now we choose an integer from 1 to 6. Let's say we choose 3. So we subtract, $6 - 3 = 3$. Now we choose an integer from 1 to 3. Let's say we choose 2. So we subtract, $3 - 2 = 1$. Now we choose an integer from 1 to 1, and of course this integer is 1. So we subtract, $1 - 1 = 0$. Having obtained 0, we stop. Our example consists of four steps (choosing an integer and subtracting).

Starting with a positive integer n, what is the expected number of steps in the Gobbling Algorithm?

Solution

Let $e(n)$ be the expected number of steps starting with the integer n. Then the first number chosen is either 1 (this happens with probability $1/n$) or not 1 (this happens with probability $(n-1)/n$). Thus, we obtain the recurrence relation

$$e(n) = \frac{1}{n}(e(n-1) + 1) + \frac{n-1}{n} e(n-1)$$

$$= e(n-1) + \frac{1}{n}, \quad e(1) = 1,$$

and it follows instantly that

$$e(n) = 1 + \frac{1}{2} + \frac{1}{3} + \cdots + \frac{1}{n}.$$

This quantity is the harmonic sum H_n (see p. 67).

Bonus: Average Number of Cycles in a Permutation

Choose a random permutation of the set $\{1, 2, 3, 4, 5, 6, 7, 8, 9, 10\}$, say,

$$\begin{pmatrix} 1 & 2 & 3 & 4 & 5 & 6 & 7 & 8 & 9 & 10 \\ 5 & 7 & 1 & 8 & 3 & 9 & 4 & 2 & 10 & 6 \end{pmatrix}.$$

Written in cycle notation, the permutation is

$$(1, 5, 3)(2, 7, 4, 8)(6, 9, 10),$$

and we see at a glance that it has three cycles. On average, how many cycles do we expect in a permutation of n elements? Surprisingly, the answer comes from the Gobbling Algorithm.

What is the relationship between the numbers produced by the Gobbling Algorithm and the cycle lengths of a random permutation? If the algorithm produces the numbers 3, 4, 3 (in the case $n = 10$), is this outcome supposed to represent all permutations of cycle type $3 + 3 + 4$? Clearly not, as the algorithm could produce these cycle lengths in a different order. Which permutations of cycle type $3 + 3 + 4$ does this outcome represent?

Actually, we can represent every permutation in a unique form that is consistent with the algorithm. The way we have written our example permutation above (on 10 elements), the smallest element in each cycle appears first, and the cycles are in order of their first elements. Let's say that a permutation written in this way is in "standard form."

Note. Normally, the order doesn't matter in a cycle type, so that, for example,

$$3 + 4 + 3 \quad \text{and} \quad 3 + 3 + 4$$

are regarded as the same. But in our situation, order matters, since different orders refer to different permutations in standard form.

What is the probability that a permutation in standard form of cycle type $3 + 4 + 3$ is chosen at random from the set of all permutations of the set $\{1, 2, 3, \ldots, 10\}$? There is only one choice for the first element of the first cycle (it must be 1), 9 choices for the second element, and 8 choices for the third element. There is only one choice for the first element of the second cycle (the smallest element not yet chosen), 6 choices for the second element, 5 choices for the third element, and 4 choices for the fourth element. There is only one choice for the first element of the third cycle (the smallest element not yet chosen), 2 choices for the second element, and 1 choice for the third element. The probability that such a permutation is chosen is therefore

$$\frac{1 \cdot 9 \cdot 8 \cdot 1 \cdot 6 \cdot 5 \cdot 4 \cdot 1 \cdot 2 \cdot 1}{10 \cdot 9 \cdot 8 \cdot 7 \cdot 6 \cdot 5 \cdot 4 \cdot 3 \cdot 2 \cdot 1} = \frac{1}{10 \cdot 7 \cdot 3}.$$

But this is the probability that the numbers 3, 4, 3 are produced by the algorithm!

Let's record our finding as a theorem.

GOBBLING ALGORITHM THEOREM. *Let $l_1, l_2, \ldots, l_k$ be positive integers whose sum is n. The probability that the Gobbling Algorithm, starting with n, produces the numbers $l_1, l_2, \ldots, l_k$ is equal to the probability that a permutation in standard form of cycle type*

$$l_1 + l_2 + \cdots + l_k$$

is chosen at random from the set of all permutations of n elements.

As a consequence of the Gobbling Algorithm Theorem, the average number of cycles in a permutation of n elements is the average number of steps in the algorithm, i.e., H_n.

We can also consider a two-dimensional version of the Gobbling Algorithm.

2-D Gobbling Algorithm

```
Let (m,n) be a an ordered pair of positive integers.
While m > 0 and n > 0, do:
  Let k₁ be a random positive integer between 1 and m,
  and k₂ a positive integer between 1 and n.
  Output (k₁,k₂).
  Let m ← m − k₁,  n ← n − k₂.
```

Given an ordered pair of positive integers (m, n), what is the expected number of steps in the 2-D Gobbling Algorithm?

2.5 Number Theory

Square Triangular Numbers

A child playing with pebbles discovers that she can place 36 pebbles in a triangle with 8 pebbles on a side or in a 6×6 square.

Thus, 36 is both a triangular number and a square number. Find another number greater than 1 that is both a triangular number and a square number.

Solution

Obviously, 0 and 1 are both triangular and square numbers, but we don't see much of a pattern looking at the numbers 0, 1, and 36. Let's introduce some variables. A triangular number is of the form $1 + 2 + \cdots + m = m(m + 1)/2$. A square number is of the form n^2. So we want a solution in integers to the equation

$$\frac{m(m+1)}{2} = n^2.$$

We already know the solutions $(m, n) = (0, 0)$, $(1, 1)$, and $(8, 6)$.

Multiplying by 8 and completing the square yields

$$(2m + 1)^2 = 8n^2 + 1.$$

Letting $x = 2m + 1$ and $y = n$, we arrive at the equivalent equation

$$x^2 - 8y^2 = 1.$$

This equation is called a Pell equation (see the Bonus). Notice that x is odd since $x = 2m + 1$, and this condition is automatically satisfied by x in the above equation, since $8y^2$ is even and the right side is odd.

Let's write down the solutions to the Pell equation that we know.

x	y	
1	0	(trivial)
3	1	(obvious)
17	6	(the child's solution)

Now, we make an educated guess that the values in the x and y columns follow a pattern. We see that $17 = 6 \cdot 3 - 1$ and $6 = 6 \cdot 1 - 0$. So it appears that the pattern, for both columns, is that you get the next number by multiplying the current number by 6 and subtracting the previous number. That is to say, the numbers in each column satisfy the recurrence relation

$$a_k = 6a_{k-1} - a_{k-2}, \quad k \geq 2.$$

If this is the case, then we can find new x and y values as follows:

$$x = 6 \cdot 17 - 3 = 99, \quad y = 6 \cdot 6 - 1 = 35.$$

Indeed, these numbers satisfy Pell's equation:

$$99^2 - 8 \cdot 35^2 = 1.$$

The corresponding values of m and n are $m = (99-1)/2 = 49$ and $n = 35$, and we have

$$49 \cdot 50/2 = 35^2 = 1225.$$

Thus, 1225 is both a triangular number and a square number.

Bonus: Pell Equations

Equations of the form $x^2 - dy^2 = 1$, where d is a positive nonsquare integer, are called Pell equations. Pell equations are named after number theorist John Pell (1611–1685), but they were studied much earlier by Brahmagupta.[9]

In our problem, we considered the Pell equation

$$x^2 - 8y^2 = 1.$$

It can be shown that all nonnegative integer solutions to this equation are given by the trivial solution $(x_0, y_0) = (1, 0)$, the obvious solution $(x_1, y_1) = (3, 1)$, and the recurrence formulas

$$x_k = 6x_{k-1} - x_{k-2},$$

$$y_k = 6y_{k-1} - y_{k-2}, \quad k \geq 2.$$

Brahmagupta left us with this challenge: "A person who can, within a year, solve $x^2 - 92y^2 = 1$ is a mathematician." Can you find a nontrivial solution to Brahmagupta's equation?

Foxy Factorial

Recall that $n!$ ("n factorial") is the product of the first n positive integers:

$$n! = 1 \cdot 2 \cdot 3 \cdot \cdots \cdot n.$$

Without performing multiplications, find the digits denoted a and b in

$$23! = 2585201ab38884976640000.$$

---

[9] Brahmagupta (598–670) made contributions in algebra, analysis, geometry, trigonometry, and number theory. His most famous work, *Brahmasphuta-siddhanta* (The Opening of the Universe), gives among other things rules for calculating with zero.

2.5 Number Theory

Solution

Do you know some divisibility rules that might help?

The sum of the given digits is 86, while the sum of all the digits must be a multiple of 9 (since 9 divides 23!). Hence $a + b = 4$ or $a + b = 13$. Also, the alternating sum of the given digits is 10, while the alternating sum of all the digits must be a multiple of 11 (since 11 divides 23!). Hence, $b - a = 1$. The only integer solution to these equations is $a = 6$ and $b = 7$.

Bonus: Divisibility Rules

In the Solution, we used two divisibility rules:

- The sum of the digits of a number leaves the same remainder upon division by 9 as the number itself.

- The alternating sum of the digits of a number leaves the same remainder upon division by 11 as the number itself.

To prove the facts, suppose that a number n has the base-10 representation

$$n = d_k \ldots d_2 d_1 d_0;$$

that is,

$$n = d_k 10^k + \cdots + d_2 10^2 + d_1 10 + d_0.$$

Reducing each term with respect to the moduli 9 and 11, we obtain

$$n \equiv d_k + \cdots + d_2 + d_1 + d_0 \pmod{9}$$

and

$$n \equiv (-1)^k d_k + \cdots + d_2 - d_1 + d_0 \pmod{11}.$$

Hence, the remainder upon dividing n by 9 (respectively, 11) is the same as the remainder upon dividing the sum of the digits (respectively, the alternating sum of digits) of n by 9 (respectively, 11). This proves the divisibility rules for 9 and 11.

Note. For any base b, these same divisibility rules work for $b - 1$ (if $b > 2$) and $b + 1$.

Always Composite

Prove that $n^4 + 4$ is a composite number (see Toolkit), for all $n \geq 2$.

Solution

Surprisingly, the polynomial $n^4 + 4$ factors. To see how, use the difference of squares factorization, $a^2 - b^2 = (a + b)(a - b)$. Thus

$$n^4 + 4 = (n^2 + 2)^2 - 4n^2 = (n^2 + 2 + 2n)(n^2 + 2 - 2n).$$

As long as the two factors, $n^2 + 2n + 2$ and $n^2 - 2n + 2$, are both at least equal to 2, their product will be a composite number. And this is easily verified: $n^2 + 2n + 2 > n^2 - 2n + 2 = n(n - 2) + 2 \geq 2$.

Bonus: The Folium of Descartes

We will show that the only point on the Folium of Descartes,[10] $x^3 + y^3 = 3xy$, with integer coordinates is $(0, 0)$. Using the factorization $x^3 + y^3 + z^3 - 3xyz = (x + y + z)(x^2 + y^2 + z^2 - xy - yz - xz)$, we let $z = 1$ and obtain $1 = x^3 + y^3 + 1 - 3xy = (x + y + 1)(x^2 + y^2 + 1 - xy - y - x)$. The only integer factorizations of 1 are $1 \cdot 1$ and $-1 \cdot -1$. In the first case, $x + y + 1 = 1$ and $x^2 + y^2 + 1 - xy - y - x = 1$, and hence $x + y = 0$, from which we obtain $(x + y)^2 = 3xy$, and so $3xy = 0$; this implies that $x = 0$ or $y = 0$ (and hence $x = y = 0$). In the second case, $x + y + 1 = -1$ and $x^2 + y^2 + 1 - xy - y - x = -1$, and hence $x + y = -2$, from which we obtain $3xy = 8$; but this is impossible since 3 does not divide 8 evenly. Hence, the only integer point on the curve is $(0, 0)$.

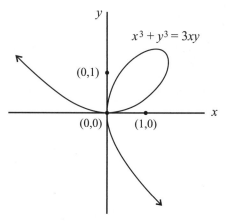

A Problem of 1's

Let
$$S = 1 + 11 + 111 + \cdots + \underbrace{1\ldots 1}_{111},$$
where the last summand has 111 decimal digits each equal to 1. What is the sum of the digits of S?

Solution

Here is a simple way to compute S:

$$9S = 9 + 99 + 999 + \cdots + \underbrace{9\ldots 9}_{111} = 10 - 1 + 10^2 - 1 + 10^3 - 1 + \cdots + 10^{111} - 1$$

$$= \underbrace{111\ldots 1}_{111}0 - 111 = \underbrace{1\ldots 1}_{112} - 112$$

$$S = (\underbrace{1\ldots 1}_{112} - 112)/9.$$

---

[10] This curve was introduced by René Descartes (1596–1650), the founder of analytic geometry. The word "folium" means "leaf" in Latin.

2.5 Number Theory

In dividing $\underbrace{1\ldots 1}_{112}$ by 9, we notice a recurring pattern due to the fact that $111111111 = 9 \cdot 12345679$. Thus, the quotient consists of blocks 123456790 (of length 9 and digital sum 37) plus "leftover" digits at the end. Since S has $111 = 9 \cdot 12 + 3$ digits, there are 12 such blocks and three leftover digits. The last four digits of S are $(1111 - 0112)/9 = 0999/9 = 0111$. Therefore, the sum of the digits of S is $12 \cdot 37 + 3 = 447$.

Bonus: A Googol

The number S has 111 digits, so it is greater than a famous number called a *googol*. A googol is 10^{100}, i.e., 1 followed by one hundred 0's. A googol is the smallest number with 101 decimal digits. The term googol was coined in 1920 by Milton Sirotta (age nine).

A Problem of 2's

Let

$$T = \underbrace{2^{2^{2^{\cdot^{\cdot^{\cdot^{2}}}}}}}_{222},$$

where the tower of exponents consists of 222 2's. What is the unit's digit of T?

Solution

The unit's digit of powers of 2 repeats in the pattern 2, 4, 8, 6. Since the exponent of the bottom 2 in T is a multiple of 4, the unit's digit of T is 6.

Bonus: Ackermann's Function

A tower of n 2's, i.e.,

$$\underbrace{2^{2^{2^{\cdot^{\cdot^{\cdot^{2}}}}}}}_{n},$$

is a fast-growing function of n. However, much faster-growing functions exist. Define $f_1(n) = 2n$, i.e., the function that doubles each input. Define

$$f_2(n) = \underbrace{(f_1 \circ \cdots \circ f_1)}_{n}(1).$$

This is the composition of f_1 with itself n times, evaluated at 1. Thus, $f_2(n) = 2^n$, an exponential function. Similarly, define

$$f_3(n) = \underbrace{(f_2 \circ \cdots \circ f_2)}_{n}(1).$$

Thus,

$$f_3(n) = \underbrace{2^{2^{2^{\cdot^{\cdot^{\cdot^{2}}}}}}}_{n},$$

our tower of 2's function. Continuing in this manner, define

$$f_4(n) = \underbrace{(f_3 \circ \cdots \circ f_3)}_{n}(1).$$

It's difficult to describe $f_4(n)$. It is a tower of 2's, where the number of 2's is itself a tower of 2's, the number of which is a tower of 2's, etc. In general, define

$$f_k(n) = \underbrace{(f_{k-1} \circ \cdots \circ f_{k-1})}_{n}(1), \quad k \geq 2.$$

The f_k are faster and faster-growing functions of n. Ackermann's function f is the "diagonal" of this process; that is,

$$f(n) = f_n(n).$$

We can easily compute only the first three terms of Ackermann's function: $f(1) = 2$, $f(2) = 4$, $f(3) = 16$. After these, the terms become very large. For instance, $f(4)$ is a tower of 65536 2's.

Ackermann's function[11] is recursive but not primitive recursive. A function is recursive if it can be obtained as the result of a well-defined algorithm. A function is primitive recursive if it can be obtained by the usual arithmetic operations, as well as composition and recursion in one variable. The function $f(n)$ can be computed via recursion (it's easy to do so using our definition of $f_k(n)$), but not recursion in one variable.

A Problem of 3's

Let

$$U = \underbrace{3\ldots3}_{333} \times \underbrace{3\ldots3}_{333},$$

where each factor consists of 333 digits each equal to 3. What is the sum of the digits of U?

Solution

We have

$$U = \underbrace{3\ldots3}_{333} \times \underbrace{3\ldots3}_{333}$$

$$= \frac{1}{3}\underbrace{9\ldots9}_{333} \times \frac{1}{3}\underbrace{9\ldots9}_{333}$$

$$= \frac{1}{9}(10^{333} - 1)(10^{333} - 1)$$

$$= \frac{1}{9}(10^{666} - 2 \times 10^{333} + 1)$$

$$= \frac{1}{9}(\underbrace{9\ldots9}_{332}8\underbrace{0\ldots0}_{332}1)$$

$$= \underbrace{1\ldots1}_{332}0\underbrace{8\ldots8}_{332}9.$$

---
[11] Ackermann's function was developed by the logician Wilhelm Ackermann (1896–1962).

2.5 Number Theory

Hence, the sum of the digits of U is

$$332 + 8 \cdot 332 + 9 = 2997.$$

Bonus: A Square Root Algorithm

Consider a "toy version" of our problem,

$$333 \times 333 = 110889.$$

So 110889 is a perfect square. If we were given this number, how would we quickly find its square root? A neat square root algorithm is a little like long division. Write the number under a radical sign, with pairs of digits separated by commas (starting at the right):

$$\sqrt{11, 08, 89}.$$

Looking at the first pair of digits, 11, we ask, what is the largest square less than or equal to this number? It is $3^2 = 9$. Write 3 above the radical and 9 below the 11.

$$\begin{array}{c} 3 \\ \sqrt{11, 08, 89} \\ 9 \end{array}$$

Now subtract the 9 from the 11 and bring down the next two digits.

$$\begin{array}{c} 3 \\ \sqrt{11, 08, 89} \\ \underline{9} \\ 2\ 08 \end{array}$$

Double the number above the radical (the 3) and write the result to the left as a two-digit number with a blank for the unit's digit.

$$\begin{array}{r} 3 \\ \sqrt{11, 08, 89} \\ \underline{9} \\ 2\ 08 \\ 6\_ \end{array}$$

What is the largest number that we can put in the blank such that this number multiplied by the two-digit number so obtained yields a product less than or equal to 208? It is 3, since $63 \times 3 = 189$, while $64 \times 4 = 256$. So put 3 in the blank and above the radical, write the product (189) beneath the 208, subtract, and bring down the next two digits.

$$\begin{array}{r} 3\ \ 3 \\ \sqrt{11, 08, 89} \\ \underline{9} \\ 2\ 08 \\ 63 \quad \underline{1\ 89} \\ 19\ 89 \end{array}$$

Now double the number above the radical to get 66, put this number to the left as a three-digit number with a blank for the unit's digit.

$$\begin{array}{r}33\\\sqrt{11,08,89}\\9\\\hline 2\ 08\end{array}$$

$$\begin{array}{rr}&\\63&1\ 89\\\hline &19\ 89\\66\_&\end{array}$$

What is the largest number that we can put in the blank such that this number multiplied by the three-digit number so obtained yields a product less than or equal to 1989? It is 3.

$$\begin{array}{r}333\\\sqrt{11,08,89}\\9\\\hline 2\ 08\end{array}$$

$$\begin{array}{rr}63&1\ 89\\\hline &19\ 89\\663&19\ 89\\\hline &0\end{array}$$

Because we have obtained a 0 remainder, the process is over. We see that the square root of 11089 is 333.

Why does this process work? Suppose that in computing the square root of n, the current number above the radical sign is x, which represents $10^k x$, for some k. Then $(10^k x)^2 \leq n$. We put a new number, y, above the radical, so that the number above the radical represents $10^k x + 10^{k-2} y$, where $(10^k x + 10^{k-2} y)^2 \leq n$, or

$$(10^k (2x) + 10^{k-2} y) \cdot 10^{k-2} y \leq n - 10^{2k} x^2.$$

This inequality is the condition that the number represented by twice the number above the radical ($2x$) with a digit appended (y) multiplied by the number represented by the digit appended (y) is less than or equal to the current remainder ($n - 10^{2k} x^2$).

To test your understanding of the algorithm, try computing the square root of 1530169.

Fibonacci Squared

Let $\{f_n^2\}$ be the sequence of squares of the Fibonacci numbers (see Toolkit). Find a linear recurrence relation with constant coefficients for this sequence.

Solution

Start with the relations

$$f_n = f_{n-1} + f_{n-2},$$
$$f_{n-3} = f_{n-1} - f_{n-2}.$$

2.5 Number Theory

Square both equations and add:

$$f_n^2 + f_{n-3}^2 = (f_{n-1} + f_{n-2})^2 + (f_{n-1} - f_{n-2})^2$$
$$= 2f_{n-1}^2 + 2f_{n-2}^2.$$

And so we obtain the recurrence relation

$$f_n^2 = 2f_{n-1}^2 + 2f_{n-2}^2 - f_{n-3}^2, \quad n \geq 3.$$

An aha! proof of the recurrence relation comes from observing that f_{n-2} and f_{n-1} are the sides, and f_{n-3} and f_n the diagonals, of a degenerate (collapsed) parallelogram. The formula given on p. 57 yields

$$f_n^2 + f_{n-3}^2 = 2f_{n-1}^2 + 2f_{n-2}^2,$$

and transposing the f_{n-3}^2 term to the right side gives our recurrence relation.

Bonus: Powers of Fibonacci Numbers

Let's investigate the same question for kth powers of Fibonacci numbers, where k is any positive integer. We want to find a linear recurrence relation with constant coefficients for the sequence $\{f_n^k\}$. The method is to use characteristic polynomials (see Toolkit under "Fibonacci numbers").

Let's work out the $k = 2$ case first (this will reproduce our recurrence relation found in the Solution). We know that the roots of the characteristic polynomial, $x^2 - x - 1$, of the Fibonacci sequence are $\phi = (1 + \sqrt{5})/2$ and $\hat{\phi} = (1 - \sqrt{5})/2$. Thus, an explicit formula for the Fibonacci numbers is

$$f_n = c_1 \phi^n + c_2 \hat{\phi}^n, \quad n \geq 0,$$

where c_1 and c_2 are constants (we could determine the constants but we don't need them). It follows that

$$f_n^2 = c_1^2 (\phi^2)^n + 2c_1 c_2 (\phi \hat{\phi})^n + c_2^2 (\hat{\phi}^2)^n,$$

and since $\phi \hat{\phi} = -1$, the characteristic roots of the polynomial for the sequence $\{f_n^2\}$ are ϕ^2, $\hat{\phi}^2$, and -1. Hence, the characteristic polynomial for this sequence is

$$(x - \phi^2)(x - \hat{\phi}^2)(x + 1) = (x^2 - (\phi^2 + \hat{\phi}^2)x + 1)(x + 1).$$

To simplify further, we introduce the Lucas numbers L_n, which satisfy the same recurrence relation as the Fibonacci numbers but with initial values $L_0 = 2$, $L_1 = 1$. Thus, the Lucas numbers are

$$2, 1, 3, 4, 7, 11, 18, 29, 47, \ldots.$$

Using the initial values, it is easy to show that

$$L_n = \phi^n + \hat{\phi}^n, \quad n \geq 0.$$

So our formula for the characteristic polynomial in the case $k = 2$ simplifies to

$$(x^2 - L_2 x + 1)(x + 1) = (x^2 - 3x + 1)(x + 1) = x^3 - 2x^2 - 2x + 1.$$

From this we confirm the recurrence relation found in the Solution.

The case $k = 3$ is similar. By the binomial theorem, the roots of the characteristic polynomial for the sequence $\{f_n^3\}$ are ϕ^3, $\phi^2\hat{\phi} = -\phi$, $\phi\hat{\phi}^2 = -\hat{\phi}$, and $\hat{\phi}^3$. Hence, the characteristic polynomial is

$$(x - \phi^3)(x - \hat{\phi}^3)(x + \phi)(x + \hat{\phi}) = (x^2 - (\phi^3 + \hat{\phi}^3)x - 1)(x^2 + (\phi + \hat{\phi})x - 1)$$
$$= (x^2 - L_3 x - 1)(x^2 + L_1 x - 1)$$
$$= (x^2 - 4x - 1)(x^2 + x - 1)$$
$$= x^4 - 3x^3 - 6x^2 + 3x + 1.$$

This gives us a recurrence relation for the cubes of the Fibonacci numbers:

$$f_n^3 = 3f_{n-1}^3 + 6f_{n-2}^3 - 3f_{n-3}^3 - f_{n-4}^3, \quad n \geq 4.$$

Using the same method, we find the characteristic polynomial for the sequence of fourth powers of the Fibonacci numbers,

$$(x^2 - L_4 x + 1)(x^2 + L_2 x + 1)(x - 1),$$

for fifth powers,

$$(x^2 - L_5 x - 1)(x^2 + L_3 x - 1)(x^2 - L_1 x - 1),$$

for sixth powers,

$$(x^2 - L_6 x + 1)(x^2 + L_4 x + 1)(x^2 - L_2 x + 1)(x + 1),$$

for seventh powers,

$$(x^2 - L_7 x - 1)(x^2 + L_5 x - 1)(x^2 - L_3 x - 1)(x^2 + L_1 x - 1),$$

and for eighth powers,

$$(x^2 - L_8 x + 1)(x^2 + L_6 x + 1)(x^2 - L_4 x + 1)(x^2 + L_2 x + 1)(x - 1).$$

The pattern is evident now, and we can write it for all $k \geq 1$ as

$$\prod_{i=0}^{\lfloor (k-1)/2 \rfloor} (x^2 + (-1)^{i+1} L_{k-2i} x + (-1)^k) \cdot \begin{cases} 1 & \text{if } k \bmod 4 = 1, 3, \\ (x - 1) & \text{if } k \bmod 4 = 0, \\ (x + 1) & \text{if } k \bmod 4 = 2. \end{cases}$$

This formula for the characteristic polynomial, found by John Riordan, means that the sequence $\{f_n^k\}$ of kth powers of the Fibonacci numbers satisfies a linear recurrence relation of order $k + 1$ with integer coefficients.

A Delight from Pascal's Triangle

Prove this gem due to Paul Erdős[12] and George Szekeres:[13] In a row of Pascal's triangle, any two numbers aside from the 1's have a common factor.

For instance, in the sixth row,

$$1 \quad 6 \quad 15 \quad 20 \quad 15 \quad 6 \quad 1,$$

the numbers 6 and 15 have the common factor 3, the numbers 6 and 20 have the common factor 2, and the numbers 15 and 20 have the common factor 5.

Solution

Suppose that the numbers are $\binom{n}{j}$ and $\binom{n}{k}$, with $0 < j < k < n$. Can you think of an identity that relates $\binom{n}{j}$ and $\binom{n}{k}$?

We employ the "subcommittee identity" (see Bonus),

$$\binom{n}{k}\binom{k}{j} = \binom{n}{j}\binom{n-j}{k-j}.$$

Obviously, $\binom{n}{j}$ divides the right side of this equation, so it also divides the left side. However, if $\binom{n}{j}$ and $\binom{n}{k}$ were coprime, then $\binom{n}{j}$ would divide $\binom{k}{j}$, but this is impossible since $\binom{n}{j} > \binom{k}{j}$. So $\binom{n}{j}$ and $\binom{n}{k}$ are not coprime; they have a common factor.

Bonus: The Subcommittee Identity

Here is an aha! proof of the identity

$$\binom{n}{k}\binom{k}{j} = \binom{n}{j}\binom{n-j}{k-j},$$

for $0 \leq j \leq k \leq n$. The left side counts the number of ways to choose, from an organization of n people, a committee of size k and a subcommittee of size j. (First choose the committee and then choose the subcommittee.) But the right side counts the same thing. (First choose the subcommittee and then choose the committee.) Therefore, the two expressions are equal.

An Unobvious Integer

Prove that, for every integer $n \geq 1$, the quantity

$$\frac{(2^n - 1)(2^n - 2)(2^n - 2^2)(2^n - 2^3)\cdots(2^n - 2^{n-1})}{n!}$$

is an integer.

---

[12] Paul Erdős (1913–1996) was one of the most prolific mathematicians of all time. He made contributions in combinatorics, graph theory, and number theory. I highly recommend George Paul Csicsery's documentary film *N is a Number: A Portrait of Paul Erdős*, about Erdős' life and mathematics.

[13] George Szekeres (1911–2005) worked in geometric combinatorics and graph theory, as well as relativity theory. Szekeres took only one course in mathematics, as an undergraduate.

Solution

Quick calculations show that the quantity equals 1, 3, 28, and 840, for $n = 1, 2, 3$, and 4, respectively.

A good method for showing that an algebraic expression is an integer is to show that it counts something. Let's illustrate this method with a very simple problem. Prove that the product of any n consecutive integers is a multiple of $n!$. This amounts to showing that the expression

$$\frac{(k+1)(k+2)(k+3)\cdots(k+n)}{n!}$$

is an integer for all integers k and positive integers n. If one of the factors in the numerator is 0, then obviously the quantity equals 0, and there is nothing to prove. If all the factors are negative, then we may factor out a -1 from each term and obtain ± 1 times a quantity in which all the factors in the numerator are positive. So, let's assume that all the factors in the numerator are positive, i.e., $k \geq 0$. Now, this quantity is equal to the binomial coefficient (see Toolkit)

$$\binom{k+n}{n},$$

so it counts the number of combinations of n objects from a set of $k + n$ objects; hence it is an integer.

Now, let's give a counting proof for our original problem. Our quantity counts the number of bases for an n-dimensional vector space over the field $F = \{0, 1\}$ (see Toolkit).

For example, there are three bases of the 2-dimensional vector space over F. They are $\{(0, 1), (1, 0)\}$, $\{(0, 1), (1, 1)\}$, and $\{(1, 0), (1, 1)\}$.

Let's show that our quantity counts these bases. Of course, there are n vectors in a basis for an n-dimensional space. I think you will guess (correctly) that the term $n!$ in the denominator accounts for all permutations of the n basis elements (since the order of elements in a basis is unimportant). We must show that there are $(2^n - 1)(2^n - 2)(2^n - 2^2)(2^n - 2^3)\ldots(2^n - 2^{n-1})$ choices for an *ordered* set of n linearly independent vectors of length n over F. There are 2^n vectors of length n over F. Any of these vectors may be taken for the first basis vector, except for the all 0 vector. Hence, there are $2^n - 1$ choices for the first basis vector. The second basis vector may be any of the 2^n vectors of length n except for those that are multiples of the first basis vector. Since there are only two possible multipliers (0 and 1), there are $2^n - 2$ choices for the second basis vector. The third basis vector may be any of the 2^n vectors of length n except for linear combinations of the first two basis vectors. A linear combination of two vectors v and w looks like $\alpha v + \beta w$, where $\alpha, \beta \in F$. Hence, there are 2^2 linear combinations and therefore $2^n - 2^2$ choices for the third basis vector. Continuing in this manner, we obtain all the factors in the numerator, one for each vector in the ordered basis.

A different (although similar) proof is obtained by realizing that the numerator of our quantity is the number of $n \times n$ invertible matrices over the field F. This is because the rows of such matrices span the n-dimensional vector space over F and hence are sequences of n linearly independent vectors of length n, as indicated above. What about the denominator? It counts the number of permutation matrices (matrices over F with a single 1 in each row

2.5 Number Theory

and column). Since the permutation matrices constitute a subgroup of the group of matrices, by Lagrange's theorem (see Toolkit), the number of such permutations ($n!$) divides the order of the group. Hence, the ratio is an integer.

Is it true that the expression

$$\frac{\prod_{k=0}^{n-1}(b^n - b^k)}{n!}$$

is an integer for every pair of positive integers b and n? Our proofs using bases or matrices extend to the case where b is a prime power (as we may replace F with a field with any prime power number of elements). However, the proofs don't work when b is not a prime power. Fortunately, a different (more prosaic) proof works for all values of b and n.

We carry out this proof by comparing the power of each prime divisor of the numerator and denominator. A prime p divides $n!$ to the power

$$\left\lfloor \frac{n}{p} \right\rfloor + \left\lfloor \frac{n}{p^2} \right\rfloor + \cdots.$$

The reason is that there are $\lfloor n/p \rfloor$ multiples of p at most equal to n; there are $\lfloor n/p^2 \rfloor$ multiples of p^2 at most equal to n, etc. The above sum is less than or equal to

$$\frac{n}{p} + \frac{n}{p^2} + \cdots,$$

which is a geometric series with sum

$$\frac{n}{p-1}.$$

Since the power of p is an integer, it is at most

$$\left\lfloor \frac{n}{p-1} \right\rfloor.$$

We will show that the power of p that divides the numerator is at least equal to this number. Since $b^n - b^k = b^k(b^{n-k} - 1)$, there is a multiple of p in the numerator for each term $b^n - b^k$ such that $b^{n-k} \equiv 1 \pmod{p}$. If b is not a multiple of p, then, by Fermat's (little) theorem (see Toolkit), this happens whenever $p - 1 \mid n - k$. Since $n - k$ ranges from n to 1 (as k ranges from 0 to $n - 1$), there are $\lfloor n/(p-1) \rfloor$ such terms. If b is a multiple of p, then the power of p in the numerator is in general much greater than the power of p in the denominator, for it is equal to $1 + 2 + \cdots + n - 1 = n(n-1)/2$ (there is a contribution of k from the factor b^k, for $1 \le k \le n - 1$), which is greater than $\lfloor n/(p-1) \rfloor$, if $p > 2$. However, in the case $p = 2, n = 2$, the inequality goes in the wrong direction. But in this case the quantity is

$$\frac{(b^2 - 1)(b^2 - b)}{2!},$$

which is an integer since the numerator is even (it contains consecutive integer factors $b-1$ and b). Therefore, in all cases, the power to which p divides the numerator is at least equal to the power to which it divides the denominator, and we are finished.

Bonus: Fermat's (Little) Theorem

We can give an aha! proof of Fermat's (little) theorem ($a^p \equiv a \pmod{p}$, for p prime) in the case $a \geq 1$ via a counting argument. How many circular necklaces can be made with p beads from a set of a different types of beads, not using only one type of bead, where two necklaces are considered the same if we can rotate one to match the other? (We call the allowed rotations "circular symmetries.") We claim that the number of such necklaces is $(a^p - a)/p$. The reason is that, without regard to the circular symmetries, there are $a^p - a$ necklaces (we subtract the a necklaces made of only one type of bead). And since p is a prime number, there are p different rotations of each necklace (think about this!). Hence, taking into account the circular symmetries, there are $(a^p - a)/p$ different necklaces. Because the number of necklaces is an integer (our good method is at work again), we obtain the congruence $a^p \equiv a \pmod{p}$.

Magic Squares

An *order n magic square* is an $n \times n$ array of integers 1 through n^2 such that the sums along every row, column, and main diagonal are the same.

(a) Create an order 3 magic square.

(b) Create an order 9 magic square.

(c) Create an order 6 magic square.

Solution

(a) Since $1 + \cdots + 9 = 45$, the constant sum must be $45/3 = 15$. With a little trial and error we find the unique (up to rotation and reflection) magic square, as shown.

8	1	6
3	5	7
4	9	2

(b) Let's compute the constant sum for a magic square of side m. The sum of all the numbers in the square is
$$1 + \cdots + m^2.$$
From our formula found in "Forward and Backward," we can write this as
$$\frac{m^2(m^2+1)}{2}.$$
Since each of the m rows has the same sum, this sum is
$$\frac{m^2(m^2+1)}{2m} = \frac{m(m^2+1)}{2}.$$

2.5 Number Theory

We can create a magic square of order 9 by superimposing an order 3 magic square with itself. It is convenient to number rows, columns, and entries starting with 0. In this case, the constant sum of an order m magic square is $m(m^2 - 1)/2$. Suppose that $A = [a(i, j)]$ is an order m magic square and $B = [b(i, j)]$ is an order n magic square. Then we can create an order mn magic square $C = [c(i, j)]$ according to the recipe

$$c(i, j) = a(\lfloor i/n \rfloor, \lfloor j/n \rfloor) \cdot n^2 + b(i \bmod n, j \bmod n), \quad 0 \le i, j \le mn - 1.$$

This puts a magic square in the same pattern as B in each place in A, numbered accordingly.

We do a sample row sum, the sum of the elements of row 0:

$$\sum_{j=0}^{mn-1} c(0, j) = \sum_{j=0}^{mn-1} [a(0, \lfloor j/n \rfloor) \cdot n^2 + b(0, j \bmod n)]$$

$$= \frac{nm(m^2 - 1) \cdot n^2}{2} + \frac{mn(n^2 - 1)}{2}$$

$$= \frac{mn(m^2 n^2 - 1)}{2}.$$

Following our recipe, we create an order 9 magic square (with constant sum 360).

70	63	68	7	0	5	52	45	50
65	67	69	2	4	6	47	49	51
66	71	64	3	8	1	48	53	46
25	18	23	43	36	41	61	54	59
20	22	24	38	40	42	56	58	60
21	26	19	39	44	37	57	62	55
34	27	32	79	72	77	16	9	14
29	31	33	74	76	78	11	13	15
30	35	28	75	80	73	12	17	10

In order to obtain a magic square numbered 1 through 81, we would add 1 to each entry of our square.

(c) The constant for an order 6 magic square is 111. It's difficult to see how to create such a square from scratch, as we can't compose a 3×3 magic square and a 2×2 magic square (because there is no 2×2 magic square). However, we can create an order 6 magic square based on the order 3 magic square. The idea is to place a 2×2 array in each cell of the order 3 magic square, and number each set of four cells consecutively according to the pattern of the order 3 square. However, there is the question of how to number the cells of the 2×2 array. For example, if we order the cells from left to right and top to bottom, we do not obtain a magic square. Here is a numbering that works (found by trial and error or the method of the Bonus):

32	29	4	1	24	21
30	31	2	3	22	23
12	9	17	20	28	25
10	11	18	19	26	27
13	16	36	33	5	8
14	15	34	35	6	7

Bonus: Conway's LUX Method

John H. Conway invented a way, called the LUX method, to create an order $2m$ magic square, given an order m magic square, where m is an odd number greater than 1. (The order six magic square above is the case $k = 1$.) Assume that $m = 2k + 1$, where $k \geq 1$. Create an $m \times m$ array consisting of $k + 1$ rows of L's, 1 row of U's, and $k - 1$ rows of X's. Interchange the L in the middle of the array with the U below it. Below is the array with $k = 4$.

L	L	L	L	L	L	L	L	L
L	L	L	L	L	L	L	L	L
L	L	L	L	L	L	L	L	L
L	L	L	L	L	L	L	L	L
L	L	L	L	U	L	L	L	L
U	U	U	U	L	U	U	U	U
X	X	X	X	X	X	X	X	X
X	X	X	X	X	X	X	X	X
X	X	X	X	X	X	X	X	X

Each cell above is filled with a 2×2 array according to one of three patterns, L, U, and X, as shown below.

$$L: \begin{matrix} 3 & 0 \\ 1 & 2 \end{matrix} \qquad U: \begin{matrix} 0 & 0 \\ 2 & 2 \end{matrix} \qquad X: \begin{matrix} 3 & 0 \\ 2 & 1 \end{matrix}$$

We will create an order $2m$ magic square, assuming numbering beginning with 0. A cell containing number z in the order m magic square is replaced by four cells with numbers $4z$, $4z + 1$, $4z + 2$, and $4z + 3$, according to the LUX numbering. Let's do some sample computations to indicate that the row, column, and diagonal sums are correct.

2.5 Number Theory

Suppose that row 0 of an order m square has elements $x_0, \ldots, x_{m-1}$. Then the sum of the elements of row 0 in the order $2m$ square is

$$(8x_0 + 3) + \cdots + (8x_{m-1} + 3) = 8(x_0 + \cdots + x_{m-1}) + 3m$$
$$= 8\left(\frac{m^3 - m}{2}\right) + 3m$$
$$= \frac{(2m)^3 - (2m)}{2}.$$

This is the correct sum for an order $2m$ magic square.

Let's also calculate the sum of the elements of column 0. Suppose that the column 0 elements of the order m magic square are $y_0, \ldots, y_{m-1}$. Since the column contains the left half of $k+1$ L's, 1 U, and $k-1$ X's, we obtain the sum

$$8(y_0 + \cdots + y_{m-1}) + 4(k+1) + 1(1) + 2(k-1) = 8(y_0 + \cdots + y_{m-1}) + 3m,$$

which we know is correct from our previous calculation.

The other row and column sums, and the diagonal contributions are similar. Can you see why an L and U must be interchanged?

All Things Being Equal

Suppose that $2n + 1$ integers have the property that upon removing any one of them, the remaining numbers can be grouped into two sets of size n with the same sum. Show that all $2n + 1$ numbers are equal.[14]

Solution

Upon removing any number, the sum of the remaining numbers is even. It follows that all the numbers are odd or all even; that is to say, all the numbers are congruent modulo 2. If the numbers are even, divide them all by 2; if they are odd, subtract 1 from each number and divide by 2. In this way, we obtain a set of integers with the same property as before. Repeating this process, we find that the original integers are all congruent modulo 4, modulo 8, etc. This implies that the numbers are all equal.

The result also holds for rational numbers simply by scaling ("clear denominators" by multiplying by the product of all the denominators).

We can also generalize the problem in another direction. Let $k \geq 2$. Suppose that $N = kn + 1$ integers have the property that upon removing any one of them, the remaining numbers can be grouped into k sets of size n with equal sums. Then all N numbers are equal. The proof is an extension of the one given above. All the numbers are congruent modulo k. Subtracting the remainder modulo k and dividing by k, we obtain a set of integers with the same property as before. Continuing in this way, we find that the original integers are all congruent modulo k, modulo k^2, etc. Hence, they are all equal.

---

[14] This problem was given in the 1973 William Lowell Putnam Mathematical Competition.

Bonus: The Passage to Real Numbers

Consider the generalization of our problem to real numbers. Suppose that $N = 2n + 1$ real numbers have the property that upon removing any one of them, the remaining numbers can be grouped into two sets of size n with equal sums. Show that all N numbers are equal.

The idea is that the $2n + 1$ real numbers generate an abelian (commutative) subgroup of the (additive) group of real numbers.[15] For example, suppose that $n = 3$ and the seven numbers are

$$\frac{1}{5},\ \sqrt{2},\ 2,\ 3,\ 3,\ \pi,\ 7.$$

Of course, this set of numbers *doesn't* have the required property; we are only illustrating how the subgroup is formed. The subgroup is the set of all linear combinations of these numbers with rational coefficients, i.e., all numbers of the form

$$a + b\sqrt{2} + c\pi, \quad a, b, c \text{ rational numbers.}$$

It's easy to show that this representation is unique. Hence, the subgroup is isomorphic to the direct product group $\mathbf{Q} \times \mathbf{Q} \times \mathbf{Q}$, where $\mathbf{Q}$ is the additive group of rational numbers. As far as our problem is concerned, the elements of this group are the only real numbers that exist.

Now, we can write each such real number in terms of "coordinates." In our little example above, (a, b, c) stands for $a + b\sqrt{2} + c\pi$. We are given that when any number is removed, the remaining $2n$ numbers can be grouped into two sets of size n with equal sums. Equal sums mean that in each coordinate, the sum of the values is the same. Thus we have, in each coordinate, the rational number version of the result that we have already proved! Hence, in each coordinate, all the rational values are equal. Therefore, all the real numbers are equal.

The real-number generalization of the $kn + 1$ problem is proved in just the same way.

2.6 Combinatorics

Now I Know My ABC's

An unusual dictionary contains as entries all permutations of the 26 letters $A, B, C, \ldots, Z$. Of course, the entries are in alphabetical order.

(a) What is the next "word" after

$$ABCDEFGHIJKLMNOPQZYXWVUTSR\ ?$$

(b) How many "words" are between

$$AZBCDEFGHIJKLMNOPQRSTUVWXY \quad \text{and}$$
$$ZABCDEFGHIJKLMNOPQRSTUVWXY\ ?$$

---

[15] Gary Martin used this idea to solve a generalization of the problem in which the "numbers" are elements of an abelian group (see the May 1988 issue of *The American Mathematical Monthly*).

2.6 Combinatorics

Solution

(a) Look at the longest decreasing (anti-alphabetical order) string that ends the word. This is ZYXWVUTSR. To find the next entry, the letter to the left of this string, Q, is interchanged with the first letter greater than Q that occurs in the decreasing string; this is R. Then the decreasing string that ends the word is reversed. Thus, the next entry is ABCDEFGHIJKLMNOPRQSTUVWXYZ.

(b) From the example in (a), we see that in going from one entry to the next, one letter is advanced and the string to the right of that letter is reversed (after switching the letter with the next highest available letter). In order for the kth letter to advance, all permutations of the following $26 - k$ letters must occur. This means that the entry number of a word is given by

$$1 + \sum_{k=1}^{26} \alpha_k (n - k)!,$$

where α_k is the number of letters to the right of the kth letter in the word that precede the kth letter alphabetically. Thus, the entry number of

AZBCDEFGHIJKLMNOPQRSTUVWXY

is $1 + 24 \cdot 24!$, while the entry number of

ZABCDEFGHIJKLMNOPQRSTUVWXY

is $1 + 25 \cdot 25!$. Hence, the number of entries between the two words is $(1 + 25 \cdot 25!) - (1 + 24 \cdot 24!) - 1 = 372,889,489,441,676,903,055,359,999$.

Bonus: An Instant Identity

Suppose that the dictionary uses n letters and hence has $n!$ entries. Then our analysis yields the identity

$$n! = 1 + (n-1)(n-1)! + \cdots + 2 \cdot 2! + 1 \cdot 1!.$$

Packing Animals in a Box

Can you put the five 4-square "animals"

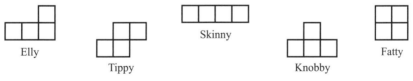

Elly, Tippy, Skinny, Knobby, Fatty

into the 4 × 5 box

without overlap? (The figures may be rotated and flipped over if necessary.)

Note. I composed this problem in 1982, when I took a class on game theory from Frank Harary (1921–2005) at the University of Michigan. Professor Harary gave the figures their whimsical names and proposed tic-tac-toe-like games in which two players attempt to make patterns of O's and X's in the shapes of the figures. The figures were independently introduced by Solomon Golomb, who called them "polyominoes."

Solution: Many people who try to solve this problem experiment with various arrangements without success and wonder why it isn't possible. They may eventually come up with valid explanations for why the figures don't fit. However, the explanations tend to be involved and wordy, with several cases to consider based on, say, which way Skinny is oriented. In fact, a succinct explanation captures the gist of the situation rather neatly.

Shade the squares of the box in a checkerboard pattern.

There are ten gray squares and ten white squares. But when the animals are put into the box, each requires two gray squares and two white squares, except for Knobby, which requires three gray squares and one white square, or one white square and three gray squares.

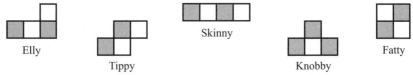

Altogether the animals require nine gray squares and 11 white squares, or 11 gray squares and nine white squares. So the animals don't fit in the box.

This is the essence of an aha! solution: it is a revelation that illuminates a situation, eliminating long, drawn-out explanations.

Bonus: Packing 3-D Animals

A 3-dimensional version of the packing problem is worth considering. There are seven 4-cube animals in 3-D:

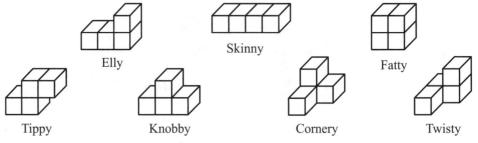

We can imagine the figures as made out of wooden blocks. The 3-D figures have the same names as the 2-D figures, in addition to the two genuinely new figures which I call Cornery and Twisty. Since these figures are composed of 28 cubes, it is natural to ask whether they can be packed into a $2 \times 2 \times 7$ box.

2.6 Combinatorics

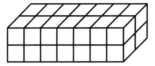

See if you can accomplish the packing. I found the following type of diagram useful for keeping track of what I was doing.

	K	K	K			

		K				

The diagram shows the top and bottom layers of the $2 \times 2 \times 7$ box. I put a Knobby (K) in the box, with three of its cubes in the top layer and one in the bottom layer.

Did you find a solution? Okay, the packing is possible, and here is one way to do it.

C	K	K	K	E	E	E
T	T	S	S	S	S	W

C	C	K	F	F	W	E
C	T	T	F	F	W	W

The key is: Tippy (T), Knobby (K), Elly (E), Skinny (S), Fatty (F), Cornery (C), Twisty (W).

I admit that when I first posed the 3-D puzzle, I tried unsuccessfully to find a packing and then spent a good deal of time trying to prove that it was impossible. Fortunately, I put forth a little more effort in the positive direction and solved it. This is typical of mathematical problem-solving. We sometimes spend time working in the positive direction (trying to show that something is possible) as well as in the negative direction (trying to prove that it is impossible). We hedge our bets and hope to come to a conclusion one way or the other.

Linear Bumper Cars

Seven particles travel up and down on a straight vertical line bounded by two endpoints. Initially, they all move upward with the same speed. When a particle strikes another particle, or reaches either endpoint, its direction of motion changes while its speed remains constant. How many particle–particle collisions occur before the particles again occupy their original positions and are moving upward?

Solution

A good idea is to start with the simplest situation that shows what's going on in the problem. Suppose that there is only one particle moving up and down on the line segment. What does its path look like over time?

The "triangular wave function" (a) shows the path of the particle as a function of time. Superimposing seven such graphs (b), we obtain the paths of all seven particles.

(a)

(b)

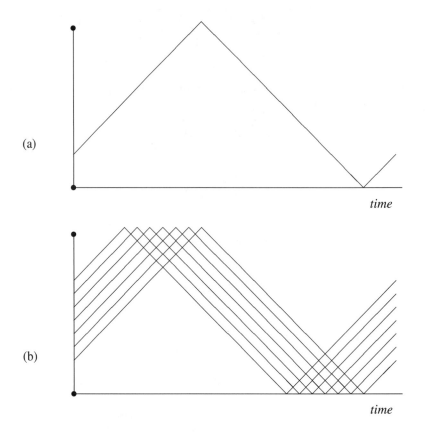

time

Strictly speaking, (b) isn't the true picture of our seven-particle system, because when two particles collide they don't pass through each other (they strike each other and reverse direction). However, the graph of two particles striking each other and reversing direction *looks exactly the same* as the graph of two particles passing through each other. That is, the pictures of the paths are the same. The only difference is which particle is on which path. Since the pictures look the same, we may as well use our picture (b) to solve the problem. We can see that every particle crosses paths twice with every other particle. Thus, the total number of crossings (and hence the total number of collisions) is $7 \times 6 = 42$ (this counts each collision twice, which is correct).

In our solution, a two-dimensional picture is worth a thousand words.

Bonus: Linear Elastic Collisions

Suppose that two particles (think of billiard balls) are moving on a straight line. One has mass m_1 and velocity v_1, the other has mass m_2 and velocity v_2. Suppose that the particles collide, undergoing either a head-on collision (if they are moving toward each other), or a bump (if they are moving in the same direction with different speeds). As a result of the collision, there is a change in the particles' velocities. The new velocities, v_1' and v_2', may be calculated by using the principles of conservation of momentum and energy. (We

2.6 Combinatorics

assume that the collision is elastic, meaning that no energy is lost.) Thus

$$m_1v_1 + m_2v_2 = m_1v_1' + m_2v_2',$$

$$\frac{1}{2}m_1v_1^2 + \frac{1}{2}m_2v_2^2 = \frac{1}{2}m_1v_1'^2 + \frac{1}{2}m_2v_2'^2.$$

Combining like terms, we obtain

$$m_1(v_1 - v_1') = m_2(v_2' - v_2),$$
$$m_1(v_1 - v_1')(v_1 + v_1') = m_2(v_2' - v_2)(v_2' + v_2).$$

Dividing the second equation by the first yields

$$v_1 + v_1' = v_2 + v_2',$$

or

$$v_1' - v_2' = -(v_1 - v_2).$$

Hence, the relative velocities of the two particles stay constant (except for a change in sign).

Solving the equations

$$v_1 + v_1' = v_2 + v_2',$$

$$m_1v_1 + m_2v_2 = m_1v_1' + m_2v_2',$$

we arrive at equations for the new velocities:

$$v_1' = \frac{m_1 - m_2}{m_1 + m_2}v_1 + \frac{2m_2}{m_1 + m_2}v_2,$$

$$v_2' = \frac{2m_1}{m_1 + m_2}v_1 + \frac{m_2 - m_1}{m_1 + m_2}v_2.$$

After the particles collide, they will not collide again. But suppose that they could. What would the new velocities v_1'' and v_2'' be? If we perform the above operations again, with v_1 and v_2 replaced by v_1' and v_2', respectively, then we find that (surprisingly) $v_1'' \doteq v_1$ and $v_2'' = v_2$. That is to say, after two collisions the particles regain their original velocities. This result is easy to show if we adopt the reference frame in which the origin is the center of mass; that is,

$$0 \equiv \frac{m}{m+n}v_1 + \frac{n}{m+n}v_2.$$

In this case, we obtain the equations $m_1v_1 + m_2v_2 = 0$, $m_1v_1' + m_2v_2' = 0$, and $v_1 + v_1' = v_2 + v_2'$, which imply that $v_1' = -v_1$ and $v_2' = -v_2$, and hence $v_1'' = v_1$ and $v_2'' = v_2$.

The hypothetical situation of two collisions can be realized if we change "linear" to "circular", so that the particles move around a frictionless circular track (with the '+' direction counter-clockwise and the '−' direction clockwise). After two head-on collisions (or bumps, if the particles are moving in the same sense), the particles are back to their original signed speeds.

I Scream Aha!

An ice-cream parlor has a special where you can order two vanilla ice-cream sundaes for the price of one. You may choose up to four toppings (repetitions allowed) for each sundae from a set of 12 toppings (cherries, nuts, chocolate syrup, etc.). The toppings for the two sundaes may be the same or different. An example of a valid order is

{cherries, chocolate syrup}   and   {cherries, cherries, nuts, nuts}.

A sign claims that altogether there are 1,048,576 different orders. You immediately refute this claim. How do you know?

Solution

Notice that 1,048,576 is a power of 2 (namely, 2^{20}). However, we will show that the number of orders *cannot* be a power of 2.

Suppose that there are x ways to order an individual sundae, according to the conditions of the special. It follows that the number of orders of two sundaes is $\binom{x+1}{2}$. A quick way to see this is to think of one more way to order a sundae, called "duplicate." When you order two sundaes, you choose them from the set of orders of one sundae plus the possibility of "duplicate."

Now we observe that $\binom{x+1}{2} = (x+1)x/2$, the product of two consecutive integers divided by 2. Since one of the two consecutive integers must be odd, the product is not a power of 2, because a power of 2 has no odd (proper) factors. This contradiction proves that the claimed number of orders is incorrect.

Bonus: The Correct Number of Orders

We know from the Solution that the number of orders of two sundaes is $(x+1)x/2$, where x is the number of orders of one sundae. How do we determine x? The number of toppings can be 0, 1, 2, 3, or 4. Let's add an extra topping to the list called "blank." When "blank" is chosen, nothing is added to the sundae. For example, the toppings selection

{cherries, chocolate syrup, blank, blank}

is the same as the selection

{cherries, chocolate syrup}.

Now we can assume that exactly four toppings are chosen (from the new set of 13 toppings). So the problem becomes, how many ways may we select four things (allowing repetitions) from a set of 13 things? This is a basic problem in combinatorics, called a *distribution problem*. The number of ways of selecting k things (repetitions allowed) from a set of n things is given by the formula $\binom{k+n-1}{k}$. Applying the formula to our problem we obtain $x = \binom{4+13-1}{4} = \binom{16}{4} = 1820$. Therefore, the number of orders of two sundaes is $1821 \cdot 1820/2 = 1{,}657{,}110$.

To explain the formula $\binom{k+n-1}{k}$ for the number of ways of selecting k things (repetitions allowed) from a set of n things, think of the things to be selected as represented by k copies of the symbol O, arranged in a row. The set of O's is partitioned into n subsets by $n-1$ vertical lines, |. The subsets represent the selections. Hence, we have $k+n-1$ items in a row, and we must choose k of them to be O's (the others being |'s). The number of ways to make these choices is $\binom{k+n-1}{k}$.

2.6 Combinatorics

As an illustration, suppose that we order one sundae with exactly four toppings (repetitions allowed) from a set of three toppings (cherries, nuts, chocolate syrup). The formula says that the number of orders is $\binom{4+3-1}{4} = \binom{6}{4} = 15$, and that each order corresponds to a placement of four O's and two $|$'s in a row. For example, the pattern of symbols for the order

$$\{\text{cherries, cherries, nuts, nuts}\}$$

is

$$O\ O\ |\ O\ O\ |.$$

The two O's to the left of the first $|$ signify that two toppings of cherries are chosen; the two O's between the two $|$'s signify that two toppings of nuts are chosen; and the absence of O's to the right of the second $|$ signifies that no chocolate syrup topping is chosen. To test your understanding, list all 15 orders and the corresponding patterns of symbols that represent them.

Lines Dividing the Plane

Into how many regions is the plane divided by n lines in general position (no two lines parallel, no three lines intersecting at a point)? For example, in the picture below, four lines divide the plane into 11 regions.

Solution

The book [11], by Ross Honsberger, shows solutions to this problem by recurrence relations and by Euler's formula (see Toolkit).

The result can also be seen "directly" using the combinatorial meaning of binomial coefficients. Add a line below all the line intersection points (the dotted line in the picture below). Note that none of the lines is horizontal. (If any line is horizontal, just rotate the picture a little.) Think of the dotted line as "cutting off" the configuration of lines; it doesn't change the number of regions bounded by solid lines. Now every region has a lowest point (i.e., a point with a smallest y-value) except the left-most region. Moreover, each region's lowest point is the intersection of either two solid lines or a solid line and the dotted line.

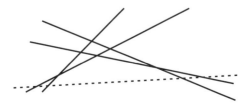

Conversely, each such intersection is the lowest point of exactly one region. Hence, the number of regions is

$$1 + n + \binom{n}{2} = 1 + n + \frac{n(n-1)}{2} = \frac{n^2 + n + 2}{2}.$$

Bonus: Partitioning Space

Into how many regions is 3-dimensional space divided by n planes in general position (no two planes parallel, no three planes intersecting in a line)? You can prove that the number of regions is

$$1 + n + \binom{n}{2} + \binom{n}{3}.$$

By analogy with the planar case, put a new plane below all the intersections of three planes. Tilt the configuration (if necessary) so that no plane is parallel to the xy-plane. Now every region has a lowest point (i.e., a point with a smallest z-value) except one. The intersection of every three planes is the lowest point of exactly one region. This counts all the regions except the ones "cut off" by the new plane. There are

$$1 + n + \binom{n}{2}$$

of these regions. So altogether there are

$$1 + n + \binom{n}{2} + \binom{n}{3} \text{ regions.}$$

In d-dimensional space, the maximum number of regions determined by n hyperplanes, i.e., $(d-1)$-dimensional planes, is

$$\binom{n}{0} + \binom{n}{1} + \binom{n}{2} + \cdots + \binom{n}{d}.$$

A Number that Counts

Prove that $\dfrac{100!}{50! \, 2^{50}}$ is an integer.

Solution

The integer has 79 digits, so we may not want to compute it by hand. There is an easy way, however, to see that the number *is* an integer.

A standard technique for showing that a number is an integer is to show that it counts something (recall "An Unobvious Integer"). In this case, the expression counts the ways of pairing 100 people into 50 pairs. The reason is that the numerator, 100!, counts the ways of permuting 100 people. Suppose that once the people are permuted, we take the first two people to be the first pair, the second two to be the second pair, etc. We have overcounted by a factor of 50! (all the ways of rearranging the pairs) and by a factor of 2^{50} (all the ways of switching the two people within a pair). Dividing by these factors results in the correct count. Therefore, the quantity is an integer.

2.6 Combinatorics

Bonus: Other Large Integers

By our counting technique, the expression

$$\frac{(2n)!}{n!2^n}$$

is an integer for all $n \geq 1$. It is the number of ways of pairing $2n$ people into n pairs. More generally, the expression

$$\frac{(mn)!}{n!(m!)^n},$$

for $m \geq 1$ and $n \geq 1$, counts the ways of grouping mn people into n groups of size m, and hence is an integer. You can amaze your friends by claiming (and proving) that

$$\frac{5300!}{100!(53!)^{100}}$$

is an integer. You certainly wouldn't want to write the number, as it has 10,319 digits.

A Broken Odometer

Alice and Bill are beginning a month-long bicycle trip when they notice that Alice's trip odometer is broken. At each mile, Alice's odometer advances the digit in the same place as the digit that Bill's odometer advances, but the digits to the right of that digit (if any) do not advance. Thus, the odometer readings for the first 25 miles of their trip are:

Alice's odometer	Bill's odometer
001	001
002	002
003	003
004	004
005	005
006	006
007	007
008	008
009	009
019	010
010	011
011	012
012	013
013	014
014	015
015	016
016	017
017	018
018	019
028	020
029	021
020	022
021	023
022	024
023	025

At the end of their trip, Bill asks Alice if she wants to know the correct mileage. Alice replies, "I can figure it out myself from my odometer." Alice's odometer reads 347. How does Alice calculate the correct mileage?

Solution

Alice explains that the rule is to add starting at the left, and take the remainder upon division by 10, to find each digit in turn. So the hundreds place is 3, the tens place is $3 + 4 = 7$, and the units place is $3 + 4 + 7 = 14 \equiv 4 \pmod{10}$. Hence the correct mileage is 374.

Why does the rule work? Notice that in the normal operation of a counter (e.g., an odometer), only one digit advances at each step of the count (any digits to the right of the advancing digit are set to 0). Hence, each step amounts to adding a vector of the form $(0, 0, \ldots, 0, 1, 1, \ldots, 1)$ to the count vector and reducing modulo 10. With the broken odometer, only one digit advances, and the digits to the right of the advancing digit do not change. This amounts to adding a vector of the form $(0, 0, \ldots, 0, 1, 0, \ldots, 0)$ to the count vector. Applying Alice's rule to the vector $(0, 0, \ldots, 0, 1, 0, \ldots, 0)$, we obtain the vector $(0, 0, \ldots, 0, 1, 1, \ldots, 1)$. Therefore, since the rule is linear, the correspondence is preserved at each step of the count.

Bonus: Finite Derivative and Integral

The sequence of numbers given by the broken odometer is called a Gray code.[16] The connection between the normal count and the Gray code count is an example of a finite derivative (going from normal count to Gray code count) and a finite integral (going from Gray code count to normal count). The derivative operator takes successive differences of numbers in the reading, starting at the left. For example, if the correct odometer reading is 457, then the broken odometer reading would be 412. The derivative operator turns the vector $(0, 0, \ldots, 0, 1, 1, \ldots, 1)$ into the vector $(0, 0, \ldots, 0, 1, 0, \ldots, 0)$. The integral operator, which takes successive sums of numbers starting at the left, performs the transformation in the reverse direction.

How Many Matrices?

The 4×4 matrix below has the properties that (1) all entries are nonnegative integers, (2) there are at most two positive entries per line (row or column), and (3) every line sum is 3.

$$\begin{bmatrix} 0 & 1 & 0 & 2 \\ 1 & 2 & 0 & 0 \\ 2 & 0 & 0 & 1 \\ 0 & 0 & 3 & 0 \end{bmatrix}$$

How many such 4×4 matrices are there?

---

[16] Gray codes, named after their inventor, physicist Frank Gray, are used in some mechanical coding devices.

2.6 Combinatorics

Solution

Our solution was found by David Callan. The matrix above can be written as

$$\begin{bmatrix} 0 & 1 & 0 & 0 \\ 1 & 0 & 0 & 0 \\ 0 & 0 & 0 & 1 \\ 0 & 0 & 1 & 0 \end{bmatrix} + \begin{bmatrix} 0 & 0 & 0 & 2 \\ 0 & 2 & 0 & 0 \\ 2 & 0 & 0 & 0 \\ 0 & 0 & 2 & 0 \end{bmatrix},$$

i.e., as a permutation matrix plus twice a permutation matrix. (A permutation matrix is a square matrix with a single 1 in each row and column and the rest of the entries 0.) In fact, any matrix that has properties (1), (2), and (3) can be written in the form $P + 2Q$, where P and Q are order 4 permutation matrices. Simply write each '3' as '1 + 2' and split the matrix up as an all 0's and 1's matrix plus an all 0's and 2's matrix. Conversely, every matrix of the form $P + 2Q$, where P and Q are order 4 permutation matrices, has properties (1), (2), and (3).

Therefore, since there are 4! permutation matrices of order 4, the number of matrices satisfying our conditions is $4!^2 = 576$ (the number of pairs of order 4 permutation matrices).

Using the same reasoning, we see that there are $n!^2$ such matrices of order n.

Bonus: Arbitrary Line Sums

Surprisingly, if we alter property (3) so that the line sums are all 2, then the problem of counting the matrices is more difficult. (If, instead, we require all the line sums to be 1, then the matrices are permutation matrices and so there are $n!$ of them.)

Let a_n be the number of matrices of order n satisfying (1) and (2), and with line sums all equal to 2. Then the numbers a_n are given by the sequence

$$1, 3, 21, 282, 6210, 202410, 9135630, 545007960, 41514583320, \ldots.$$

It is known that the a_n satisfy the recurrence relation

$$a_n = n^2 a_{n-1} - \frac{1}{2} n(n-1)^2 a_{n-2}, \quad n \geq 3.$$

However, no simple direct formula for a_n is known.

Richard Stanley has shown [20, p. 151] that if the line sums are all equal to an arbitrary positive integer r, then the number of such matrices of order n is $n!^2$ times the coefficient of x^n in the power series expansion of

$$(1-x)^{(1-r)/2} e^{(3-r)x/2}.$$

The $r = 3$ case is miraculously nice!

Lots of Permutations

Prove that given any collection of $n^d - n^{d-1} + 1$ distinct d-tuples from the set $S = \{1, 2, \ldots, n\}$, there exist n of the d-tuples which, in each coordinate, are a permutation of S. Show that the result is not true for a collection of d-tuples of size $n^d - n^{d-1}$.

Solution

Partition the n^d d-tuples into n^{d-1} pairwise disjoint subsets of size n each, so that each of these subsets consists of n points which in each coordinate are a permutation of $\{1, 2, 3, \ldots, n\}$. (A simple way to do it: for each vector v with 1 in the first coordinate, take all its cyclic shifts, that is, $v, v + (1, 1, \ldots, 1), v + (2, 2, \ldots, 2)$, etc., where addition is taken modulo n.) Now, to destroy all these subsets, we have to omit at least one element from each of them. Hence, we have to omit at least n^{d-1} vectors, showing that $n^d - n^{d-1}$ is tight; it is the largest possible cardinality of a set of vectors with no forbidden configuration.

The collection of d-tuples in which the first coordinate is not equal to 1 does not have the desired property, and there are $n^d - n^{d-1}$ such d-tuples. Hence, the result is not true for $n^d - n^{d-1}$ d-tuples.

Bonus: Integer Averages

P. Erdős, A. Ginzburg, and A. Ziv proved that given $2n - 1$ integers, some n have an integer average. For example, with $n = 5$, among the nine integers

$$7,\ 8,\ 3,\ -4,\ 7,\ 0,\ 10,\ 1,\ 4$$

there are five with an integer average, e.g., 3, −4, 0, 10, and 1.

The value $2n - 1$ is best-possible, as a collection of $n - 1$ 0's and $n - 1$ 1's (a set of size $2n - 2$) does not have the desired property.

We will give a proof by contradiction that draws on Fermat's (little) theorem and the multinomial theorem (see Toolkit). Observe that we only have to prove the result for n prime. For suppose that we have proved the result for a and b. If $n = ab$, then given any $2n - 1$ integers we can successively find subsets of size a with an integer average. How many such subsets can we find? Upon identifying the first subset of size a, we have $2ab - 1 - a = 2a(b-1) - 1$ integers left. Upon identifying the second such subset, we have $2a(b-1) - 1 - a = 2a(b-2) - 1$ integers left. Altogether, we can determine $2b - 1$ such subsets, as $2ab - 1 = a(2b - 1) + a - 1$. By hypothesis, some b of the integer averages have an integer average, and hence the corresponding ab integers have an integer average.

Now, assume that p is prime and the collection $x_1, x_2, \ldots, x_{2p-1}$ of integers has no subset of size p with an integer average. Form the following sum over of all possible subsets of size p:

$$\sum_{1 \leq k_1 < k_2 < \ldots < k_p \leq 2p-1} (x_{k_1} + x_{k_2} + \cdots + x_{k_p})^{p-1}.$$

By Fermat's (little) theorem, each summand is congruent to 1 modulo p. Hence

$$\sum_{1 \leq k_1 < k_2 < \ldots < k_p \leq 2p-1} 1 \equiv \binom{2p-1}{p} \pmod{p}.$$

The binomial coefficient *is not* a multiple of p (since the numerator contains one factor of p and so does the denominator). The contradiction will arise from showing that the sum *is* a multiple of p.

To see this, note that each monomial contains fewer than p different variables. It follows that the number of occurrences of a given monomial is $\binom{p}{i}$, where $0 < i < p$. Such a binomial coefficient is divisible by p since there is a factor of p in the numerator and none in the denominator.

2.6 Combinatorics

Even Steven and Oddball

In the game Even Steven, two players alternately take one or two stones from a pile of n stones (n an odd number). The object is to take, in total, an even number of stones. Someone must win, because whenever an odd number is the sum of two numbers, one of the numbers is even and the other odd. Who should win Even Steven, the first player or the second player? The game Oddball is identical to Even Steven except that the winner is the player who takes an odd number of stones. Who wins Oddball?

Solution

The pattern depends on the remainder of n when divided by 4.

n	Even Steven	Oddball
1 (mod 4)	2nd player wins	1st player wins
3 (mod 4)	1st player wins	2nd player wins

For $n = 1$ or $n = 3$, the values of the table are trivial. If $n = 1$, then the first player takes one stone and immediately loses (in Even Steven) or immediately wins (in Oddball). If $n = 3$, then the first player wins Even Steven by taking two stones (and the second player then takes one stone and the game is over). As for Oddball (when $n = 3$), the second player has a winning strategy: if the first player takes two stones, then the second player takes one (and wins); if the first player takes one stone, then the second player takes one stone and the first player is forced to take another stone (hence the second player wins).

Larger values of n always reduce to the values 1 or 3. Assume that the table is correct for all odd $n < N$. Now, if $N \equiv 1 \pmod{4}$, then the first player wins Oddball by taking two stones, for the game is reduced to the game where $N - 2 \equiv 3 \pmod{4}$, in which the second player (formerly known as the first player) wins. If $N \equiv 3 \pmod{4}$, then the first player wins Even Steven by taking two stones and reducing the game to the game where $N - 2 \equiv 1 \pmod{4}$ with a second player win (that is, the original first player wins). The other two cases ($N \equiv 1 \pmod{4}$ in Even Steven and $N \equiv 3 \pmod{4}$ in Oddball) are dealt with a little differently. Whatever move the first player makes, the second player mimics, so that after a pair of moves, either four stones have been taken (two by each player), and the game is represented by exactly the same entry in our table, with the second player winning, or two stones have been taken (one by each player). If two stones have been taken, then the column in the table that represents the game is changed (from Even Steven to Oddball or *vice versa*) and the row is changed (from 1 mod 4 to 3 mod 4 or *vice versa*). In any event, we go from one entry in the table (in which the second player wins) to the diagonally opposite entry (in which the second player also wins).

The above explanation is a little involved. A simpler explanation comes (as is so often the case) from expanding on the problem. In fact, the expansion is warranted. For what would happen in practical play of the games? The players might not make the best possible moves. For instance, if $n = 21$, then the first player should win Oddball by taking two stones as a first move. But suppose that the first player instead takes one stone. This leaves 20 stones in the pile and our table doesn't say what to do. Perhaps the second player has a chance to win, or perhaps the first player's move doesn't matter and the first player is still winning. Let's find out what's going on in general by allowing any number of stones (even or odd) in the initial pile. If the initial number is even, then both players will end up with an even number of stones or both an odd number of stones; hence, we need to change the goal of the game. Let's say that in Even Steven (starting with any number of stones), the first player wins if he/she takes an even number of stones, and the second player wins if this does not happen. Similarly, the first player wins Oddball (starting with any number of stones) if he/she takes an odd number of stones, and the second player wins if this does not happen.

Now we can quickly generate a table showing who wins both generalized games.

n	Even Steven	Oddball
1	2nd player wins	1st player wins
2	1st player wins	1st player wins
3	1st player wins	2nd player wins
4	1st player wins	2nd player wins
5	2nd player wins	1st player wins
6	1st player wins	1st player wins

We've already determined the $n = 1$ row. For $n = 2$, the first player wins Even Steven by taking two stones, and the first player wins Oddball by taking one stone (and forcing the second player to take one stone). We can generate the next two rows ($n = 3$ and $n = 4$) from the first two rows. In general, we determine any row from the previous two rows. Let's do this for $n = 3$. (We have already determined the $n = 3$ row, but let's do it again to see the pattern.) In the game Even Steven, the first player wants to take an even number of stones and so does the second player. Thus, the first player looks at the previous two rows ($n = 2$ and $n = 1$) of the Even Steven column to see if there is a second player win (because he/she will be in effect the second player after the current move). There *is* a second player win in the $n = 1$ row, so the first player takes two stones. Now the first player (newly called the second player) wins. A similar situation occurs in Oddball (for $n = 3$). Both players want to take an odd number of stones. Thus, the first player looks at the previous two rows ($n = 2$ and $n = 1$) of the Oddball column to see if there is a second player win. There is not a second player win. Hence, whichever move the first player makes (taking one or two stones), he/she cannot win. Therefore, the second player wins. The calculation works similarly for any odd row. The first player always looks at the previous two rows in the same column to see if there is a second player win. If there is, then he/she moves into that position. If there isn't, then he/she cannot force a win and may as well move randomly. The procedure for an even row is a little different. Let's illustrate it for $n = 4$. In Even Steven, the first player wants to take an even number of stones and the second player wants to prevent this; hence the second player wants to take an odd number of stones. Thus, the first player looks at the previous two rows ($n = 3$ and $n = 2$) in the Oddball game to see if there is a second player win (for he/she will shortly be the second player, trying to prevent the opponent from taking an odd number of stones). There is a second player win (in the $n = 3$ row). Hence, the first player moves accordingly (taking one stone). The case of Oddball is dealt with similarly. The first player wants to take an odd number of stones, while the second player wants to take an even number of stones. Thus, the first player looks at the previous two rows ($n = 3$ and $n = 2$) in the Even Steven column to see if there is a second player win. There is not a second player win, so the first player cannot force a win and just moves randomly.

2.6 Combinatorics

Try to use the above procedure to determine row $n = 5$. Remember that in the case of an odd value of n, you look at the previous two rows in the *same* column to see if there is a second player win. Did you get the same answers as in the table above? Try it for $n = 6$, as well. Remember that for even rows, you look at the previous two rows in the *opposite* column. Did you get the same values as in the table above?

In our table, we see that rows $n = 5$ and $n = 6$ are the same as rows $n = 1$ and $n = 2$, respectively. Since our procedure is defined at every step only in terms of the previous two rows, we are guaranteed that the pattern we see in the first four rows is repeated forever. To play a game, the first player calculates n modulo 4 and decides whether he/she can force a win by using our procedure. If so, the player plays accordingly. If not, he/she plays randomly. The second player does the same thing. If either player makes a mistake (has a win but plays an erroneous move), then the other player determines what he/she needs to win and plays accordingly.

Where is the aha! solution in this long discussion? I would say that it is found in the above table. Once we understand how it is generated, we understand everything about the games.

Bonus: Variations on the Games

It's surprising that in the repeating period of four rows, there is not an even distribution between first player and second player wins (there are five first player wins and three second player wins). To gain some perspective about this, we can change the games to allow the players to remove one, two, or three stones at each turn. You might think that this would make no difference to the outcomes, as one and three have the same parity, but this is wrong. The resulting game table is quite different, with a period of eight rows. You can use a modification of our procedure (looking at the previous *three* rows at each step) to generate the table. More generally, consider the modified Even Steven and Oddball games in which the players may take 1 through k stones on each turn, where k is a positive integer. Show that the resulting table has a period of $2k + 2$ rows if $n = 2k$ (an even integer) and $4k$ rows if $n = 2k - 1$ (an odd integer).

Higher-Dimensional Tic-Tac-Toe

Tic-tac-toe is played on a 3×3 board. Two players, Oh and Ex, alternately write naughts O and crosses X in unoccupied cells of the board, with each player's objective to make "three-in-a-row" in his or her symbol. The three cells may be connected horizontally, vertically, or diagonally. Two versions of the board are shown below.

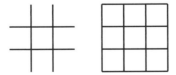

It is well known that with best-possible play by both players, tic-tac-toe is a draw. Neither player can achieve the goal of three-in-a-row.

In our problem, we are interested in counting the number of winning lines. It's easy to do this for the 3×3 board. There are three horizontal lines, three vertical lines, and two diagonal lines; so altogether there are eight winning lines.

If tic-tac-toe is played on a $3 \times 3 \times 3$ cube, so that the goal is to complete three cubes in a line, how many winning lines are there?

Solution

A very simple solution uses "outside the box" thinking. Put the $3 \times 3 \times 3$ game cube in the center of a $5 \times 5 \times 5$ cube. Every winning line in the game cube can be extended by one cube in both directions. The extension cubes lie in the "margin," i.e., the part of the $5 \times 5 \times 5$ cube that is not in the $3 \times 3 \times 3$ game cube. Hence, the number of winning lines is the number of pairs of cube in the margin, i.e.,

$$(5^3 - 3^3)/2 = 49.$$

The same technique allows us to count the number of winning lines in n^k tic-tac-toe, that is, tic-tac-toe played on a k-dimensional board of side length n. The goal is to complete n cells in a line. Place the n^k game board in the center of an $(n+2)^k$ board. Extend each winning line by one cell in both directions. The extension cells are in the margin of size $(n+2)^k - n^k$. Hence, the number of winning lines is

$$((n+2)^k - n^k)/2.$$

Bonus: A General Drawing Strategy

Who should win tic-tac-toe on a higher-dimensional board? A basic tenet of game theory is that in a "taking game" (where both players take objects to try to achieve a goal), only the first player can win given best-possible play. The reason is that if the second player had a win, then the first player could copy the winning strategy and "get there first." So the real question is whether the first player has a forced win or the second player can force a draw.

Paul Erdős and John Selfridge proved a general theorem about when the second player can draw in a taking game. Suppose that a finite collection $\{A_i\}$ of n-element subsets of a finite universe are given. Two players take turns claiming elements of the universe. The first player to claim all the elements of some A_i is the winner. If neither player achieves this, then the game is a draw.

THE ERDŐS–SELFRIDGE THEOREM. If the number of winning sets is less than 2^{n-1}, then the second player has a forced draw. On the other hand, there exist games with 2^{n-1} winning sets in which the first player can force a win.

We give a sketch of the proof. Consider the game based on the universe $\{z, x_1, y_1, x_2, y_2, \ldots, x_{n-1}, y_{n-1}\}$, where the winning sets consist of z together with exactly one of x_i, y_i, for $1 \leq i \leq n-1$. Then the winning sets have n elements each and there are 2^{n-1} of them. The first player can win this game by first choosing z and then choosing x_i if the second player chooses y_i, or vice versa, for each i. This establishes the second half of the theorem.

Now we must show that if the number of winning sets is less than 2^{n-1}, then the second player has a drawing strategy. Suppose that there are fewer than 2^{n-1} winning sets in the

2.6 Combinatorics

game. Remember that these sets have size n. At any point when it is the second player's turn to play, we define the "danger" of the position to indicate how precarious the position is for the second player. Sets from which the second player has already selected some points represent no danger to the second player (since they cannot be completed by the first player); so we discount such sets. We say that the danger of any other set is 2^m, where the first player has chosen m of the elements. The danger of the position is the total danger of all the sets. If the first player occupied all n elements of the set, then the danger of that set would be 2^n. We will give a strategy by which the second player can prevent this. For each element in the collection, we define the "score" of that element to be the sum of the danger of all the sets that contain that element. The second player chooses the element with the highest score. Now the first player gets a turn. We will argue that the result of the second player's turn and the first player's turn cannot increase the danger of the position. The second player's move removes from consideration all sets that contain the chosen element. The first player's move doubles the score of each element in the collection of sets containing the chosen element. Since the second player chose a point with maximal score, the net result of these two moves is to decrease the danger (or leave it unchanged). Note that the danger of the position after the first player's first move is less than $2^{n-1} \cdot 2^1 = 2^n$. Since the danger never increases, the first player can never force a position where the danger is 2^n. Hence the first player cannot win.

Now we apply the Erdős–Selfridge theorem to tic-tac-toe. Consider the game played on an n^k board. As long as the number of winning lines is less than 2^{n-1}, then Ex can force a draw. From our formula for the number of winning lines, the criterion is

$$((n+2)^k - n^k)/2 < 2^{n-1}.$$

Numerical investigation shows, for instance, that Ex draws for $k = 5$ and $n = 22$, that is, tic-tac-toe on a 5-dimensional board of side length 22.

The Spice of Life

A teacher has 25 students. She wishes to make five groups of five students each. The groups should be changed every day. How does she do this over a several-day period in such a way that the number of times that two students are in the same group is minimized?

Solution

Any particular student will work with four students each day, so he/she will work with 24 students in six days. If these 24 students are all different, then the student will work with each classmate exactly once during the six days. So the best possible schedule would be a six-day schedule in which every student works with every other student exactly once. We will describe a simple way to produce such a schedule.

Arrange the 25 students in five rows of five students each. Label the rows 0, 1, 2, 3, and 4, and the columns 0, 1, 2, 3, and 4. Also, number the days 0, 1, 2, 3, and 4. On day 0, let the columns be the groups, so that students in the same column work in the same group. On day 1, shift every student in row 1 to the right one place (the student at the end of the row gets moved to the beginning of the row); shift every student in row 2 to the right two places (again, using "wrap-around" here and throughout); shift every student in row 3 to the right three places; and shift every student in row 4 to the right four places; don't change row 0. The columns of this new arrangement constitute day 1's groups. Repeat this process on days 2, 3, and 4. On day 5, let the rows be the groups.

This procedure allows every student to work with every other student. Here is why: If two students are in the same row, then they work together on day 5. Suppose that two students are in different rows, with the second student r rows down ($r \neq 0$) and c columns to the right of the first student. In the rightward march, the two students are separated by c columns, $r + c$ columns, $2r + c$ columns, $3r + c$ columns, and $4r + c$ columns on days 0, 1, 2, 3, and 4, respectively (considering "wrap-around"). These numbers constitute a *complete residue system* modulo 5; that is, the numbers c, $r + c$, $2r + c$, $3r + c$, and $4r + c$ are equal to the numbers 0, 1, 2, 3, and 4 (modulo 5), in some order. Since one of the numbers is 0, the two students are in the same group on some day. To see that we have a complete residue system, notice that the congruence $xr + c \equiv y \pmod 5$ has the solution $x \equiv r^{-1}(y - c) \pmod 5$. We just need to prove that $r^{-1} \pmod 5$ exists for each $r \neq 0$. Well, 2 and 3 are multiplicative inverses modulo 5, and 1 and 4 are multiplicative self-inverses modulo 5. Hence, there is a solution x to the congruence.

Bonus: A Finite Field Method

Would our method work for a class of 16 students (arranged in four rows of four students each)? No, it would not. The flaw is that the congruence $xr + c \equiv y \pmod 4$ does not always have a solution. For example, the congruence $2x + 1 \equiv 0 \pmod 4$ has no solution, because an odd number is not a multiple of 4.

However, a schedule based on a finite field works in the case of 16 students. In fact, the method will work for any number n^2 of students, where n is a prime power, for a field of order n exists in these cases.

Let F be a field of n elements. Arrange the students in an $n \times n$ array, with the rows and columns indexed by the n field elements:

$$\{(x, y): x, y \in F\}.$$

We will construct an $(n + 1)$-day schedule in which every student works with every other student exactly once.

We index the days by the elements of F, together with ∞ for the last day. For each $k \in F$, on day k the groups are the sets of points on the "lines"

$$y = kx + b, \quad b \in F.$$

On day ∞, the groups are the columns of the array, that is,

$$\{(k, y): y \in F\}, \quad k \in F.$$

With this system, each student works with every other student exactly once, because there is exactly one intersection of any two lines.

Let's use our method to construct the schedule for a class of 16 students. First, we need a field of 4 elements. To construct this field, we require an irreducible polynomial of degree

2.6 Combinatorics

2 over the field of two elements, 0 and 1. (This means that the polynomial doesn't factor into linear polynomials whose coefficients are 0 or 1.) The polynomial $x^2 + x + 1$ works. Suppose that θ is a root of this polynomial, i.e.,

$$\theta^2 + \theta + 1 = 0.$$

Then, since all coefficients are 0 or 1, we have

$$\theta^2 = \theta + 1.$$

Multiplying by θ, we obtain

$$\theta^3 = \theta^2 + \theta = \theta + 1 + \theta = 2\theta + 1 = 1.$$

Since $\theta^3 = 1$, there are only three different powers of θ, namely, $\theta^0 = 1$, θ, and $\theta^2 = \theta + 1$. These three elements are the nonzero elements of our field. When we include 0, we have our field of four elements. For convenience, let's relabel θ as a and θ^2 as b. It's easy to figure out addition and multiplication in the field from the two rules, $a + b = 1$ and $ab = 1$.

Now we can solve the "Spice of Life" problem for a class of 16 students, according to the recipe given above. For ease in reading, let's indicate an ordered pair of field elements (x, y) by xy. Here is the five-day schedule:

Day 0

$$\{00, 10, a0, b0\}$$
$$\{01, 11, a1, b1\}$$
$$\{0a, 1a, aa, ba\}$$
$$\{0b, 1b, ab, bb\}$$

Day 1

$$\{00, 11, aa, bb\}$$
$$\{01, 10, ab, ba\}$$
$$\{0a, 1b, a0, b1\}$$
$$\{0b, 1a, a1, b0\}$$

Day a

$$\{00, 1a, ab, b1\}$$
$$\{01, 1b, a0, b0\}$$
$$\{0a, 10, a1, bb\}$$
$$\{0b, 11, a0, ba\}$$

Day b

$\{00, 1b, a1, ba\}$

$\{01, 1a, a0, bb\}$

$\{0a, 11, ab, b0\}$

$\{0b, 1a, aa, b1\}$

Day ∞

$\{00, 01, 0a, 0b\}$

$\{10, 11, 1a, 1b\}$

$\{20, 21, 2a, 2b\}$

$\{30, 31, 3a, 3b\}$.

Sperner's Lemma

Prove Sperner's lemma:

SPERNER'S LEMMA.[17] Let T be a triangle with vertices are labeled 1, 2, 3 (a "123 triangle"). Some new points are added to the edges and/or the interior of T. Points on edge 12 of T are labeled either 1 or 2; points on edge 13 are labeled 1 or 3; points on edge 23 are labeled 2 or 3; and points in the interior of T are labeled 1, 2, or 3. Finally, T is triangulated into smaller triangles whose vertices are the new points or the vertices of T. Then one of these smaller triangles is a 123 triangle.

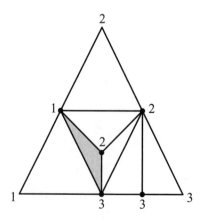

---

[17] Sperner's lemma was proved by Emanuel Sperner (1905–1980). The analysis is used in fair division algorithms.

2.6 Combinatorics

Solution

The aha! realization is that a triangle with an odd number of edges labeled 12 must be a 123 triangle. We will show that there must be such a triangle. If there were no triangle with an odd number of edges labeled 12, then the total number of edges labeled 12, over all smaller triangles, would be an even number. But we will show that this total, call it s, is in fact odd. There are two cases of 12 edges to consider: (1) those that are entirely on the 12 edge of the large triangle T, and (2) those that are not. There is an odd contribution to s from the case (1) edges. For, starting at vertex 1 on T, and moving along the edge toward vertex 2, we encounter a 12 edge only when we switch from a vertex labeled 1 to a vertex labeled 2. Since there are an odd number of such switches, there are an odd number of 12 edges. As for case (2), a 12 edge in the interior of T contributes 2 to s, as it is an edge of two different triangles. Hence, the total contribution to s from the case (2) edges is even. Therefore, accounting for an odd contribution from case (1) and an even contribution from case (2), we conclude that s is odd. This finishes the proof.

Bonus: Brouwer's Fixed-Point Theorem

Brouwer's fixed-point theorem (an important theorem of analysis), proved by Luitzen Egbertus Jan Brouwer,[18] follows easily from Sperner's lemma.

BROUWER'S FIXED-POINT THEOREM. Suppose that f is a continuous map from the unit disk $D \in \mathbf{R}^2$ into itself. Then f has a fixed point, i.e., a point $p \in D$ such that $f(p) = p$.

We demonstrate the result for a triangle (and its interior) rather than for a disk. We can do this because a disk and a triangle are homeomorphic, and hence the two sets have the same topological properties. Let the given triangle be $T = ABC$. We assign *barycentric coordinates* to all points in T (and its interior) with respect to the vertices A, B, and C. We think of A, B, and C as vectors and coordinatize each point x as $\langle x_1, x_2, x_3 \rangle$, where $x = x_1 A + x_2 B + x_3 C$, $x_1 + x_2 + x_3 = 1$, and $x_i \geq 0$, for $i = 1, 2, 3$. Notice that A, B, and C have coordinates $\langle 1, 0, 0 \rangle$, $\langle 0, 1, 0 \rangle$, and $\langle 0, 0, 1 \rangle$, respectively. Now let ABC be partitioned into many small triangles and let the vertices of these small triangles be labeled as follows: If f maps vertex $x = \langle x_1, x_2, x_3 \rangle$ to the point $x' = \langle x_1', x_2', x_3' \rangle$, then we assign x the label i if $x_i \geq x_i'$, for $i = 1, 2, 3$. Notice that A is labeled 1, B is labeled 2, and C is labeled 3, and each vertex must have some label (since the sum of the coordinates of any point is 1) but can have two or even three labels. Sperner's lemma implies the existence of a small 123 triangle T'. Repeating this process yields an infinite sequence of arbitrarily small, nested 123 triangles:

$$ABC = T \supseteq T' \supseteq T'' \supseteq T''' \supseteq \cdots.$$

Since triangles (with interiors) are compact sets, there exists a point p contained in all the triangles of this sequence. We claim that p is a fixed point of f. As p is arbitrarily close to points labeled 1, points labeled 2, and points labeled 3, it follows that p is labeled 1, 2, and 3. And any point labeled 1, 2, and 3 must be a fixed point, since the relations $x_1 \geq x_1'$, $x_2 \geq x_2'$, $x_3 \geq x_3'$, and $x_1 + x_2 + x_3 = x_1' + x_2' + x_3'$ imply that $x_1 = x_1'$, $x_2 = x_2'$, and $x_3 = x_3'$.

---

[18]Luitzen Egbertus Jan Brouwer (1881–1966) worked in analysis and topology.

An Infinite Series

Determine the sum $\sum_{n=1}^{\infty} nf_n/3^n$, where f_n is the nth Fibonacci number. Recall that $f_0 = 0$, $f_1 = 1$, and $f_n = f_{n-1} + f_{n-2}$, for $n \geq 2$.

Solution

The aha! realization is that a generating function underlies the sum. (A generating function is a power series whose coefficients form a sequence of interest.) Define

$$f(x) = \sum_{n=1}^{\infty} f_n x^n.$$

Then

$$f(x) = x + x^2 + 2x^3 + 3x^4 + 5x^5 + 8x^6 + \cdots,$$
$$xf(x) = x^2 + x^3 + 2x^4 + 3x^5 + 5x^6 + 8x^7 + \cdots,$$
$$x^2 f(x) = x^3 + x^4 + 2x^5 + 3x^6 + 5x^7 + 8x^8 + \cdots.$$

If we subtract the second and third equations from the first equation, almost everything cancels (because of the recurrence relation). Hence

$$f(x)(1 - x - x^2) = x,$$

and so

$$f(x) = \frac{x}{1 - x - x^2}.$$

In order to "manufacture" the n in our sum, we take a derivative. In this way, our sum is determined to be

$$(xf'(x))\big|_{x=1/3} = 6/5.$$

Bonus: Rational Generating Functions

Suppose that a sequence $\{a_n\}$ satisfies a recurrence relation

$$a_n = \alpha_1 a_{n-1} + \alpha_2 a_{n-2} + \cdots + \alpha_k a_{n-k}, \quad n \geq k,$$

with initial values $a_0, a_1, \ldots, a_{k-1}$, where the α_i are constants. Then the generating function

$$\sum_{n=0}^{\infty} a_n x^n$$

is a rational function of the form

$$\frac{p(x)}{1 - \alpha_1 x - \alpha_2 x^2 - \cdots - \alpha_k x^k},$$

where $p(x)$ is a polynomial of degree less than k. To obtain $p(x)$, calculate

$$(a_0 + a_1 x + \cdots + a_{k-1} x^{k-1})(1 - \alpha_1 x - \alpha_2 x^2 - \cdots - \alpha_k x^k)$$

and discard all terms of degree k or higher.

2.6 Combinatorics

As an application, let's find the generating function of the Lucas numbers L_n, which satisfy the same recurrence relation as the Fibonacci numbers, i.e., $L_n = L_{n-1} + L_{n-2}$, for $n \geq 2$, with the initial values $L_0 = 2$ and $L_1 = 1$. We calculate

$$(2+x)(1-x-x^2) = 2 - x - 3x^2 - x^3,$$

and so the generating function is

$$\frac{2-x}{1-x-x^2}.$$

Can you find the sum $\sum_{n=1}^{\infty} nL_n/3^n$?

Change for a Dollar

Suppose that we have an unlimited supply of 1-cent, 3-cent, 5-cent, and 10-cent coins. Are there more ways to make 100 cents using an even number of these coins or using an odd number of these coins?

Solution

Rather than try to count all the ways of making 100 cents, let's generate some data for making the amount n, where n is small.

n	# even ways	# odd ways	n	# even ways	# odd ways
1	0	1	11	1	8
2	1	0	12	9	1
3	0	2	13	2	10
4	2	0	14	11	2
5	0	3	15	3	13
6	4	0	16	14	4
7	0	4	17	4	15
8	5	0	18	17	5
9	0	6	19	6	18
10	7	1	20	21	7

For example, if $n = 10$, there are seven even ways ($1+1+1+1+1+1+1+1+1+1$, $1+1+1+1+1+1+1+3$, $1+1+1+1+3+3$, $1+3+3+3$, $1+1+3+5$, $1+1+1+1+1+5$, $5+5$) and one odd way (10).

We conjecture that if n is odd, there are more ways to make n with an odd number of coins than with an even number, while if n is even, there are more ways to make n with an even number of coins than with an odd number.

In order to prove our conjecture, we use the method of generating functions. Let

$$f(x) = a_0 + a_1 x + a_2 x^2 + a_3 x^3 + a_4 x^4 + a_5 x^5 + \cdots,$$

where a_n is the number of ways of making the amount n using an even number of coins minus the number of ways of making n using an odd number of coins. We also define $a_0 = 1$. Thus

$$f(x) = 1 - x + x^2 - 2x^3 + 2x^4 - 3x^5 + \cdots.$$

A little reflection shows that

$$f(x) = \frac{1}{(1+x)(1+x^3)(1+x^5)(1+x^{10})}.$$

This may be seen by examining the contribution to the generating function from each term in the denominator. For example, the term $1/(1+x^3)$ yields

$$\frac{1}{1+x^3} = 1 - x^3 + x^6 - x^9 + x^{12} - x^{15} + \cdots.$$

In the product of such terms, one of the monomials above is selected. For instance, if the $-x^9$ term is selected, this accounts for three 3-cent coins, and a contribution of -1 to the sign of a_n (which makes sense, since three is odd). The other terms make similar contributions, so that a_n is equal to the number of ways to make n using an even number of coins minus the number of ways to make n using an odd number of coins. Now, we perform a little algebra on $f(x)$ to obtain

$$f(x) = \frac{(1-x)(1-x^3)(1-x^5)}{(1-x^2)(1-x^6)(1-x^{20})}.$$

Since

$$\frac{1}{1-x^2} = 1 + x^2 + x^4 + x^6 + x^8 + x^{10} + \cdots,$$

the contribution to the generating function from $1/(1-x^2)$ consists of only even powers of x with positive coefficients. The same is true for the terms $1/(1-x^6)$ and $1/(1-x^{20})$, and hence also for the product $1/((1-x^2)(1-x^6)(1-x^{20}))$. Therefore, as far as the truth of our conjecture is concerned, the contribution from this product is "neutral" (it doesn't change the sign of a coefficient of any power of x). The numerator when expanded is a polynomial, namely, $1 - x - x^3 + x^4 - x^5 + x^6 + x^8 - x^9$, whose coefficients are positive for even powers and negative for odd powers. Therefore, the coefficients of the generating function $f(x)$ alternate in sign, with $a_n > 0$ if n is even and $a_n < 0$ if n is odd.

We have proved our conjecture, and in particular there are more ways to make 100 with an even number of coins than with an odd number.

Bonus: More Change for a Dollar

The generating function for making change using pennies, nickels, dimes, quarters, fifty-cent pieces, and whole dollars is

$$f(x) = \frac{1}{(1-x)(1-x^5)(1-x^{10})(1-x^{25})(1-x^{50})(1-x^{100})}.$$

Using a computer algebra system, we find that the coefficient of x^{100} of this generating function is 293, i.e., there are 293 ways to make change for a dollar.

Rook Paths

A chess Rook can move any number of squares horizontally or vertically on a chess board. How many different paths can a Rook travel in moving from the lower-left corner (a1) to the upper-right-corner (h8) on the board? Assume that the Rook moves right or up at every step. For example, one path is a1-c1-d1-d5-f5-f7-g7-g8-h8.

2.6 Combinatorics

Solution: It's often a good idea to generalize a problem. Let's consider Rook paths to any square on the board (with the Rook starting on a1 and moving right or up at every step). We make a table displaying the number of paths. Shifting our orientation, let's say that the upper-left entry of the table is the number of paths from a1 to a1, that is, 1. We find each new entry by adding all the entries above or to the left of the given entry. The reason is that the Rook's last move must come from one of the squares represented by these entries. So, for example, the entry in the (3, 4) position is $4 + 12 + 2 + 5 + 14 = 37$. Calculating the entry corresponding to each square of the chess board, we find that the number of paths to h8 is 470010. The dots in the table mean that we can further generalize our problem by considering arbitrarily large chess boards.

1	1	2	4	8	16	32	64	...
1	2	5	12	28	64	144	320	...
2	5	14	37	94	232	560	1328	...
4	12	37	106	289	760	1944	4864	...
8	28	94	289	838	2329	6266	16428	...
16	64	232	760	2329	6802	19149	52356	...
32	144	560	1944	6266	19149	56190	159645	...
64	320	1328	4864	16428	52356	159645	470010	...
⋮	⋮	⋮	⋮	⋮	⋮	⋮	⋮	

If we are lazy doing our sums, we can feed partial data, say, the diagonal elements, 1, 2, 14, 106, into The On-Line Encyclopedia of Integer Sequences (see "What's the Next Term?"). We find that the sequence (A051708) is

$$1, 2, 14, 106, 838, 6802, 56190, 470010, \ldots.$$

The eighth term, 470010, is the number of Rook paths in our problem.

Bonus: A Generating Function for the Diagonal

Let $f(x)$ be the generating function for the diagonal of our table, i.e.,

$$f(x) = x + 2x^2 + 14x^3 + 106x^4 + 838x^5 + 6802x^6 + 56190x^7 + 470010x^8 + \cdots.$$

Then

$$f(x) = \frac{1}{2}\left(x + \frac{x(1-x)}{\sqrt{(1-x)(1-9x)}}\right).$$

Is it obvious? Hardly. Let's prove it.

Call the elements of the above table $a(m, n)$, and set $a(m, n) = 0$ if $m = 0$ or $n = 0$. It's easy to see that for positive integers m and n, the sequence $a(m, n)$ satisfies a recurrence relation (with initial conditions):

$$a(m, n) = 2a(m, n-1) + 2a(m-1, n) - 3a(m-1, n-1), \quad m \geq 3 \text{ or } n \geq 3;$$

$$a(1, 1) = 1, \ a(1, 2) = 1, \ a(2, 1) = 1, \ a(2, 2) = 2.$$

It follows that the generating function for the doubly-infinite sequence is a rational function, namely,

$$\frac{st(1 - s - t + st)}{1 - 2s - 2t + 3st}.$$

Expanded as a power series, the monomials are of the form $a(m, n)s^m t^n$, where m and n are positive integers.

In order to get the generating function for the diagonal sequence[19], we make the change of variables $t = x/s$ (so that $st = x$). Now, to accommodate terms such as $s^5 t^7$, we allow arbitrary integer exponents for s. The exponents of x are positive integers. Thus, we represent $s^5 t^7$ as $s^{-2} x^7$.

So we obtain the generating function

$$\frac{x(1 - s - x/s + x)}{1 - 2s - 2x/s + 3x} = \frac{1}{2}\left(x + \frac{x(1 - x)s}{-2s^2 + (3x + 1)s - 2x}\right).$$

We see some of the terms of the claimed generating function. We need only concentrate on the function

$$\frac{s}{-2s^2 + (3x + 1)s - 2x},$$

which, using the quadratic formula, we write as

$$\frac{s}{-2(s - \alpha)(s - \beta)},$$

where

$$\alpha = \frac{3x + 1 - \sqrt{(1 - x)(1 - 9x)}}{4}, \quad \beta = \frac{3x + 1 + \sqrt{(1 - x)(1 - 9x)}}{4}.$$

The diagonal generating function is the coefficient of s^0, because no s occurs in it.

We put our function into partial fractions form,

$$\frac{1}{2(\beta - \alpha)}\left[\frac{\alpha}{s - \alpha} - \frac{\beta}{s - \beta}\right],$$

or

$$\frac{1}{2(\beta - \alpha)}\left[\frac{\alpha/s}{1 - (\alpha/s)} + \frac{1}{1 - (s/\beta)}\right].$$

---

[19] If a two-variable sequence satisfies a linear recurrence relation, then the generating function of its main diagonal sequence is algebraic. This statement follows from a result in the theory of functions known as Puiseux's theorem, after Victor Puiseux (1820–1883), a mathematician who worked in elliptic functions and astronomy.

2.6 Combinatorics

For $-1/9 < x < 1/9$, we expand the function by a Laurent series[20] in the annulus $|\alpha| < |s| < |\beta|$, in powers of (α/s) and (s/β), obtaining

$$\frac{1}{2(\beta - \alpha)} \left[\sum_{n=1}^{\infty} \left(\frac{\alpha}{s}\right)^n + \sum_{n=0}^{\infty} \left(\frac{s}{\beta}\right)^n \right].$$

The coefficient of s^0 is

$$\frac{1}{2(\beta - \alpha)} = \frac{1}{\sqrt{(1-x)(1-9x)}}.$$

We have found the last piece of our generating function $f(x)$.

For more information on the method we have used, see [20]. Can you find the generating function for the number of King paths from the lower-left corner to the upper-right corner of an arbitrary-size chess board? At each step, the King moves one square to the right, one square up, or both. Such paths are called *Delannoy paths*, named after Henri Delannoy (1833-1915).

Now that we have the generating function for the diagonal sequence, we can use it to obtain a recurrence relation for the sequence. Rather than work with $f(x)$ itself, it is easier to work with

$$g(x) = 2f(x) - x = \frac{x\sqrt{1-x}}{\sqrt{1-9x}}.$$

Note that f and g represent sequences satisfying the same recurrence relation but with different initial conditions. The goal is to find an equation satisfied by g and some of its derivatives, with polynomial coefficients. (We'll see why this is good shortly.) Via logarithmic differentiation, we obtain

$$\frac{g'(x)}{g(x)} = \frac{1}{x} + \frac{-\frac{1}{2}}{1-x} + \frac{\frac{9}{2}}{1-9x},$$

and hence

$$g'(x)x(1-x)(1-9x) = g(x)(1-3x)^2.$$

Sequences that satisfy such equations are called *D-finite*, meaning that the equations contain only a finite number of derivatives.

The corresponding equation for f is

$$f'(x)x(1-x)(1-9x) = f(x)(1-3x)^2 - 2x^2,$$

from which it follows that

$$[xf'(x) - f(x)] + [-10x^2 f'(x) + 6xf(x)] + [9x^3 f'(x) - 9x^2 f(x)] = -2x^2.$$

We can read off the recurrence relation for f by examining the coefficient of x^n in the generating function:

$$(n-1)a_n + (-10(n-1) + 6)a_{n-1} + (9(n-2) - 9)a_{n-2} = 0, \quad n \geq 3.$$

---

[20]Laurent series were introduced by Pierre Alphonse Laurent (1813–1854).

Therefore, we can write the recurrence formula for $\{a_n\}$ as

$$a_1 = 1;\ a_2 = 2;$$
$$a_n = \frac{(10n-16)a_{n-1} - (9n-27)a_{n-2}}{n-1}, \quad n \geq 3.$$

Surely, this is not obvious. I don't know of a counting proof of this recurrence relation. Perhaps you can find one.

3
Advanced Problems

An advanced problem may have an aha! solution. This doesn't necessarily mean that the solution is easy, only that a key step is the product of inspired thinking. You may wish to turn to the Toolkit for some mathematical terms in the statements of these problems. In some cases, the solutions require advanced techniques and concepts.

3.1 Geometry

Self-Intersecting Polygons

Given any polygon, join the midpoints of the edges (in order) to produce a new polygon. Notice that the pentagon below yields a "child" (shown with dotted lines) that is non-self-intersecting.

 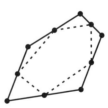

However, the pentagram below yields a child that is self-intersecting.

If we repeat this process, then clearly our first example above produces a sequence of descendant polygons that are all convex and hence all non-self-intersecting, while the second polygon above produces a sequence of descendants that are all self-intersecting.

Find a polygon that produces a sequence of descendants that are alternately self-intersecting and non-self-intersecting.[1]

Solution

We saw in "A Quadrilateral from a Quadrilateral" that the child of any quadrilateral is a parallelogram and hence all its descendants are non-self-intersecting (or the parallelogram is degenerate and all its descendants are self-intersecting). However, a pentagon exists with the required property. The aha! idea is to use the golden ratio.

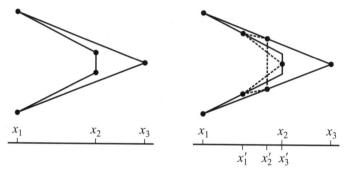

We set $(x_2 - x_1)/(x_3 - x_2) = \phi = (1 + \sqrt{5})/2$ and form a non-self-intersecting pentagon as shown above. (The points' vertical coordinates don't matter.) Then the horizontal coordinates of the child are in the same proportion as those of the original pentagon, for

$$\frac{x_2' - x_1'}{x_3' - x_2'} = \frac{(x_1 + x_3)/2 - (x_1 + x_2)/2}{x_2 - (x_1 + x_3)/2}$$

$$= \frac{x_3 - x_2}{2x_2 - x_1 - x_3}$$

$$= \frac{1}{(x_2 - x_1)/(x_3 - x_2) - 1}$$

$$= \frac{1}{\phi - 1}$$

$$= \phi.$$

It's easy to check that the "grandchild" of our pentagon is non-self-intersecting, so it follows that the descendants of the pentagon are alternately self-intersecting and non-self-intersecting.

Bonus: Non-Intersecting Transformations

Let C be a curve in the plane that doesn't intersect itself. For simplicity, suppose that C is piecewise linear. Under what conditions does there exist a similar copy of C arbitrarily

---
[1] This problem was posed and solved by E. R. Berlekamp, E. N. Gilbert, and F. W. Sinden.

3.1 Geometry

close to C without intersecting C? Of course, we must define our terms, but the picture below conveys the meaning.

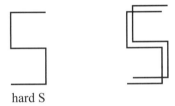

The curve C' is obtained from C by translating downward a small distance. In general, let's allow any similarity transformation. This includes scaling and rotation, as well as translation (but let's exclude reflection).

An example of a figure that doesn't work is a "hard S," made of five line segments, as in the picture below. There is no way to put a copy of the hard S close to the original without overlapping.

hard S

I don't know what properties of C allow for such a transformation, but the resolution of the problem may depend on "bottlenecks" in the curve. A bottleneck is a local minimum of the distance between two points on the curve, such as between the two points labeled b on the curve below.

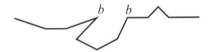

Regular Simplices

Suppose that $v_1, \ldots, v_n$ are points in $\mathbf{R}^d$ such that $|v_i - v_j| = 1$ for each distinct pair v_i, v_j. Show that $n \leq d + 1$.

Note. A set of $d+1$ points in $\mathbf{R}^d$ all the same distance apart, and the line segments joining them, is called a *regular simplex*. For example, an equilateral triangle is a regular simplex in $\mathbf{R}^2$ and a regular tetrahedron is a regular simplex in $\mathbf{R}^3$.

Solution

Let $w_i = v_i - v_n$, for $i = 1, \ldots, n-1$. We will show that the w_i are linearly independent and therefore $n - 1 \leq d$. Note that $|w_i| = 1$, for each i, and hence $w_i \cdot w_i = 1$. Also, for $i, j < n$ and $i \neq j$, note that v_i, v_j, and v_n are the vertices of an equilateral triangle, and hence $w_i \cdot w_j = 1/2$. Suppose that

$$\alpha_1 w_1 + \cdots + \alpha_{n-1} w_{n-1} = 0.$$

Taking the dot product of both sides of this equation with w_1 yields the relation

$$\alpha_1 + \frac{1}{2}\alpha_2 + \cdots + \frac{1}{2}\alpha_i + \cdots + \frac{1}{2}\alpha_{n-1} = 0,$$

and, for $2 \leq i \leq n-1$, taking the dot product with w_i yields

$$\frac{1}{2}\alpha_1 + \frac{1}{2}\alpha_2 + \cdots + \frac{1}{2}\alpha_{i-1} + \alpha_i + \frac{1}{2}\alpha_{i+1} + \cdots + \frac{1}{2}\alpha_{n-1} = 0.$$

Subtracting these equations implies that $\frac{1}{2}\alpha_1 - \frac{1}{2}\alpha_i = 0$, i.e., $\alpha_i = \alpha_1$. Hence, all the α_i are equal and thus all equal to 0. Therefore, the w_i are linearly independent and the result follows.

A regular simplex exists for each d and is easy to construct. The d points $(1, 0, 0, \ldots, 0)$, $(0, 1, 0, \ldots, 0)$, $\ldots$, $(0, 0, \ldots, 0, 1)$ in $\mathbf{R}^d$ are all distance $\sqrt{2}$ apart. We need only determine a new point that is at distance $\sqrt{2}$ from all of them. Let the new point be $(a, a, \ldots, a)$, with

$$(a-1)^2 + (d-1)a^2 = 2.$$

Using the quadratic formula, we obtain two solutions: $a = (1 \pm \sqrt{1+d})/d$. Thus, there are two choices for the new point, one with positive coordinates and one with negative coordinates.

Bonus: Lattice Point Vertices

We saw in the Solution that the maximum number of points in $\mathbf{R}^d$ all the same distance apart is $d + 1$. Can we find such a collection of points with integer coordinates? We call a point with integer coordinates a *lattice point*.

We see from the Solution that if $d + 1$ is a perfect square, then a is a rational number. Let $d + 1 = s^2$, where s is a positive integer. Then

$$a = \frac{1 \pm s}{s^2 - 1} = \frac{1}{s-1} \text{ or } -\frac{1}{s+1}.$$

By scaling, we have two sets of $d + 1$ lattice points:

$$\begin{array}{ll} (s-1, 0, \ldots, 0) & (s+1, 0, \ldots, 0) \\ (0, s-1, 0, \ldots, 0) & (0, s+1, 0, \ldots, 0) \\ \quad\vdots & \quad\vdots \\ (0, \ldots, 0, s-1) & (0, \ldots, 0, s+1) \\ (1, \ldots, 1) & (-1, \ldots, -1). \end{array}$$

For example, in $\mathbf{R}^3$, we obtain two regular tetrahedra with integer coordinates:

$$\begin{array}{ll} (1, 0, 0) & (3, 0, 0) \\ (0, 1, 0) & (0, 3, 0) \\ (0, 0, 1) & (0, 0, 3) \\ (1, 1, 1) & (-1, -1, -1). \end{array}$$

We have shown that a simplex in $\mathbf{R}^d$ with lattice point vertices exists if $d + 1$ is a perfect square. Next, we show that if d is even, then such a simplex exists only if $d + 1$ is a perfect square.

3.1 Geometry

Consider the case $d = 2$. We will show that there is no equilateral triangle in $\mathbf{R}^2$ with integer coordinates. The method is to calculate the area of the triangle in two different ways, one yielding a rational number and the other yielding an irrational number. This contradiction will show that there is no such equilateral triangle. By elementary geometry, the area of an equilateral triangle with side length s is $s^2\sqrt{3}/4$. Suppose that the triangle has lattice point vertices. Then s^2 is an integer, and we conclude that the area is irrational. However, the area is also given by the determinant formula

$$\frac{1}{2}|\det[v_1 - v_3, v_2 - v_3]|,$$

where the rows of the 2×2 matrix are the vectors $v_1 - v_3$ and $v_2 - v_3$. But this is a rational number!

For the general case, we need the formula for the volume of a regular simplex in $\mathbf{R}^d$ with edges of length s. The formula is

$$\frac{\sqrt{d+1}}{d!2^{d/2}}s^d.$$

For example, the volume of a regular tetrahedron ($d = 3$) with edge length $s = 1$ is $\sqrt{2}/12$, a formula that we found in "Volume of a Tetrahedron."

We will prove the general volume formula by induction, for edges of length $\sqrt{2}$; in this case, the volume is

$$\frac{\sqrt{d+1}}{d!}.$$

The general formula follows by scaling. The formula is obvious if $d = 1$. Assuming the formula for $d - 1$, we show that it holds for d. We must multiply the volume formula for $d - 1$ by

$$\frac{\sqrt{d+1}}{\sqrt{d}} \cdot \frac{1}{d}.$$

What accounts for this factor? Take d vertices to be standard basis vectors in $\mathbf{R}^d$. The centroid of this collection of points is $(1/d, \ldots, 1/d)$. The "height" of the simplex is the distance from the centroid to the point $(a, \ldots, a)$, with $a = (1 \pm \sqrt{1+d})/d$, that we found in the Solution. This distance is $\sqrt{d+1}/\sqrt{d}$. The other term in the factor, $1/d$, is the d-dimensional version of the multiplication factor $1/2$ used to obtain the area of a triangle in $\mathbf{R}^2$. This completes the induction and the proof of the volume formula.

Now, assuming that d is even, the denominator of the volume formula is an integer. If the vertices of the regular simplex are lattice points, then s^d is also an integer. Hence, the volume is rational if and only if $d + 1$ is a perfect square. As in the $d = 2$ case, the determinant formula for volume,

$$\frac{1}{d!}|\det[v_1 - v_d, v_2 - v_d, \ldots, v_{d-1} - v_d]|,$$

shows that the volume is a rational number (when d is even). This completes our argument.

A theorem of Isaac Jacob Schoenberg[2] says that there exists a regular simplex with lattice point vertices in $\mathbf{R}^d$ if and only if (1) $d + 1$ is an odd square, (2) $d \equiv 3 \pmod{4}$, or (3) $d \equiv 1 \pmod{4}$ and $d + 1$ is the sum of two squares.

---

[2] Isaac Jacob Schoenberg (1903–1990) is best known for his invention of the spline method of approximation.

$n^2 + 1$ Closed Intervals

Given $n^2 + 1$ closed finite intervals on the real line, show that some $n + 1$ of them have a point in common or some $n + 1$ of them are pairwise disjoint.

Solution

Let's define an order on the intervals. We say that interval $[a, b]$ is *less than* interval $[c, d]$ if and only if $b < c$. Note that two intervals are comparable if and only if they are disjoint. The aha! step is to assign to each interval the length of the longest "chain" that begins with that interval, where a chain is a sequence of intervals each of which lies entirely to the left of the next. If any of these numbers is $n + 1$ or more, then the intervals making up that chain are pairwise disjoint and we are finished. So let's assume that all the numbers are at most n. It follows by the pigeonhole principle (see Toolkit) that some number occurs at least $n + 1$ times. But no intervals assigned to the same number can be in a chain together (an interval to the left would have a greater number than an interval to the right). So every pair of these $n + 1$ intervals intersect. The maximum left endpoint of the intervals (or, just as well, the minimum right endpoint) is a point common to the $n + 1$ intervals.

Bonus: Dilworth's Lemma

A generalization of the Problem is worth considering.

DILWORTH'S LEMMA.[3] Let P be a partial order on a set of $mn + 1$ elements. Then P contains a chain of length $m + 1$ or an antichain of size $n + 1$.

A few definitions are needed. A *partial order* on a set S is a binary relation $\leq$ on S that satisfies the following conditions:

- (reflexivity) $a \leq a$, for all $a \in S$;

- (anti-symmetry) $a \leq b$ implies that $b \not\leq a$, for all $a, b \in S$ and $a \neq b$;

- (transitivity) $a \leq b$ and $b \leq c$ imply that $a \leq c$, for all $a, b, c \in S$.

In a partially ordered set, two elements a and b are *comparable* if $a \leq b$ or $b \leq a$. Otherwise, the elements are *noncomparable*. A *chain* in a partially ordered set is a collection of elements in which every pair is comparable. An antichain is a collection of elements no two of which are comparable.

The proof of Dilworth's Lemma is similar to the proof given in the Solution. The solution to our problem follows immediately from Dilworth's Lemma, as a chain is a collection of pairwise disjoint intervals and an anti-chain is a collection of intervals that share a point. The fact that the relation in the Solution isn't reflexive is unimportant.

You may wonder whether every partial order on a finite set is equivalent to an order given by intervals on the real line. This is not the case, and there is a simple counterexample: $\{(a, a), (b, b), (a, b), (c, c), (d, d), (c, d)\}$. This partial order, called $2 \oplus 2$, cannot be represented by intervals on the real line (try it!). However, this is the only type of counterexample, as asserted by the following theorem of Peter Fishburn.

---
[3] This formula is credited to R. P. Dilworth (1914–1993), a pioneer in lattice theory.

THEOREM. A partial order on a finite set is an interval order if and only if it does not contain a copy of $2 \oplus 2$.

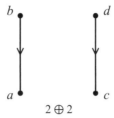

$2 \oplus 2$

3.2 Probability

1,000,000 Coin Flips

Suppose that we flip 1,000,000 fair coins and obtain X heads. Which event is more likely,

(A) $X = 500,000$ or (B) $0 \leq X \leq 495,000$?

Solution

The X in our problem is called a binomial random variable. The distribution of a binomial random variable X is denoted by $B(n, p)$ to indicate that X is the sum of n independent random variables each equal to 1 with probability p and 0 with probability $1 - p$. Here the variables are the coin flips, with $n = 1,000,000$ and $p = 1/2$. For $0 \leq k \leq n$, the probability that $X = k$ is given by the formula

$$\Pr\{X = k\} = \binom{n}{k} p^k (1-p)^{n-k}.$$

For event (A), we have

$$\Pr\{X = 500,000\} = \binom{1,000,000}{500,000} \left(\frac{1}{2}\right)^{1,000,000}.$$

An approximation is afforded by Stirling's approximation (see Toolkit):

$$n! \sim n^n e^{-n} \sqrt{2\pi n}.$$

It follows that

$$\binom{n}{\lfloor n/2 \rfloor} \approx \frac{2^n}{\sqrt{\pi n/2}}$$

and hence

$$\binom{1,000,000}{500,000} \left(\frac{1}{2}\right)^{1,000,000} \approx \frac{1}{\sqrt{\pi \, 500,000}} \approx 0.000798.$$

For event (B), we have

$$\Pr\{0 \leq X \leq 495{,}000\} = \sum_{k=0}^{495{,}000} \binom{1{,}000{,}000}{k} \left(\frac{1}{2}\right)^{1{,}000{,}000}.$$

If $X \sim B(n, p)$, then X has mean $\mu = np$ and standard deviation $\sigma = \sqrt{npq}$. In our problem, $\mu = 500{,}000$ and $\sigma = 500$. The cut-off point for X is 10 standard deviations from the mean. This implies that the probability of event (B) is extremely small. In order to estimate the size, we establish a bound on the tail of the binomial distribution. For $0 < p < 1$, define

$$H(p) = -p \log p - q \log q,$$

where $q = 1 - p$ and the logarithms are base 2. This is the *entropy function* of Information Theory (see Bonus).

CLAIM: If $0 < x < 1/2$, then

$$\sum_{k=0}^{\lfloor nx \rfloor} \binom{n}{k} < 2^{nH(x)}.$$

Here is a proof. Let $y = 1 - x$. Then

$$1 = (x + y)^n$$

$$= \sum_{k=0}^{n} \binom{n}{k} x^k y^{n-k}$$

$$> \sum_{k=0}^{\lfloor nx \rfloor} \binom{n}{k} (x/y)^k y^n$$

$$> \sum_{k=0}^{\lfloor nx \rfloor} \binom{n}{k} (x/y)^{nx} y^n$$

$$= \sum_{k=0}^{\lfloor nx \rfloor} \binom{n}{k} x^{nx} y^{ny}.$$

Therefore,

$$\sum_{k=0}^{\lfloor nx \rfloor} \binom{n}{k} < x^{-nx} y^{-ny} = 2^{nH(x)}.$$

This concludes the proof.

So, if $X \sim B(n, \frac{1}{2})$, then

$$\Pr\{0 \leq X \leq \lfloor nx \rfloor\} = 2^{-n} \sum_{k=0}^{\lfloor nx \rfloor} \binom{n}{k}$$

$$< 2^{-n} \cdot 2^{nH(x)}$$

$$= 2^{-n(1-H(x))}$$

$$= 2^{-nC(x)},$$

3.2 Probability

where $C(x) = 1 - H(x)$ is called the *capacity function*. (We will see more about this function in the Bonus for the next Problem.) If $x = 0.495$, then $2^{-nC(x)} \approx (0.99995000041665)^n$. Therefore
$$\Pr\{0 \leq X \leq 495{,}000\} < (0.9999500005)^{1{,}000{,}000}.$$

This bound is an extremely small number, approximately 1.92×10^{-22}. Comparing our calculations for the probabilities of events (A) and (B), we see that the probability of event (A) is astronomically larger (by a factor of at least 4.14×10^{18}) than the probability of event (B).

Bonus: The Entropy Function

Suppose that X is a random variable that takes values $x_1, \ldots, x_n$ with probabilities $p_1, \ldots, p_n$, respectively, where each $p_i \geq 0$ and $\sum_{i=1}^{n} p_i = 1$. If we learn that $X = x_i$, then we say that we receive $-\log_2 p_i$ *bits* of information. Suppose that X takes one value after another (independently), according to its probability distribution. Then the average amount of information obtained per value of X is

$$H(X) = -\sum_{i=1}^{n} p_i \log_2 p_i \ \text{ bits.}$$

We call $H(X)$ the *entropy* of X. The concept of entropy was introduced by Ludwig Boltzmann (1844–1906) in 1896, and Claude Shannon (1914–2001) was the first to apply it to information sources (in 1948).

We write log for $\log_2$ and set $0 \log 0 = \lim_{x \to 0^+} x \log x = 0$.

If X takes two values, x_1 and x_2, occurring with probabilities p and q, respectively, then

$$H(X) = -p \log p - q \log q.$$

We also denote this expression by $H(p)$. Below is the graph of this function.

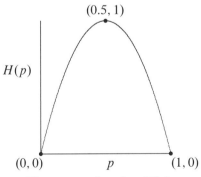

The entropy function $H(p)$.

The importance of the entropy function in Information Theory is due to two theorems of Shannon, one on coding an information source and the other on coding over a noisy channel. We discuss Shannon's First Theorem here.

Shannon's First Theorem says that there is an encoding of the values of X, with strings 0 and 1, no string a prefix of another, so that the average length of the code words is within

1 bit of $H(X)$. Furthermore, the average length of the code words is always at least equal to $H(X)$.

For example, suppose that X is a random variable with the following distribution:

$$X = \begin{pmatrix} x_1 & x_2 & x_3 & x_4 & x_5 & x_6 & x_7 & x_8 \\ 0.2 & 0.2 & 0.1 & 0.1 & 0.1 & 0.1 & 0.1 & 0.1 \end{pmatrix}.$$

Then $H(X) = 2(-0.2 \log 0.2) + 6(-0.1 \log 0.1) \approx 2.9$. A suitable code as guaranteed by Shannon's First Theorem is

$$\begin{pmatrix} x_1 & x_2 & x_3 & x_4 & x_5 & x_6 & x_7 & x_8 \\ 0.2 & 0.2 & 0.1 & 0.1 & 0.1 & 0.1 & 0.1 & 0.1 \\ 100 & 101 & 0000 & 0001 & 0010 & 0011 & 0100 & 0101 \end{pmatrix}.$$

The average length of this code is

$$0.2(3) + 0.2(3) + 0.1(4) + 0.1(4) + 0.1(4) + 0.1(4) + 0.1(4) + 0.1(4) = 3.6,$$

which is greater than, but within 1 bit of, $H(X)$.

Bits of Luck

You have the opportunity to play a game in which you can make a fortune. You possess a biased coin with probability 0.53 of landing heads and probability 0.47 of landing tails. You start the game with \$1, and once a day—every day for the rest of your life—you may bet any fraction of your current amount on the event that the coin lands heads. If the coin lands heads, you win an amount equal to the amount bet; if the coin lands tails, you lose the amount bet. What fraction of your current amount should you bet, on a day-to-day basis, in order to maximize your long-term profit?

Solution

Let's solve the problem in general, where the coin lands heads with probability p and tails with probability q, with $p + q = 1$ and $p > q > 0$.

It's clear that you don't want to bet all of your current amount, because you could lose and then you wouldn't be able to play anymore. But what fraction should you bet? Think outside the box. Of course, it doesn't make sense to bet on the coin landing tails (even if you could) because the coin is biased toward heads. However, let's imagine the possibility of placing bets on heads *and* tails.

If you divide your entire fortune on the two types of bets, some of the bets will cancel out. A hunch is that you should bet p of your current amount on heads and q of your current amount on tails. In this case, q of the bets will cancel out, leaving a remainder of $p - q$ of your amount bet on heads. Specifically, with $p = 0.53$, this is equivalent to betting $0.53 - 0.47 = 0.06$ of your current amount on heads.

Let's prove that the hunch is correct. After n days, the expected number of heads is pn and the expected number of tails is qn. Suppose that you bet λ_1 of your amount on heads and λ_2 of your amount on tails, where $\lambda_1 + \lambda_2 = 1$. We wish to show that the best choice

3.2 Probability

is $\lambda_1 = p$ and $\lambda_2 = q$. Each occurrence of heads yields a return of $2\lambda_1$ of your amount, while each occurrence of tails yields a return of $2\lambda_2$ of your amount. Hence, after n days, the expected value of your amount is

$$(2\lambda_1)^{np}(2\lambda_2)^{nq}.$$

We can write this as an exponential function of n, with 2 as the base:

$$2^{cn}, \quad \text{where } c = 1 + p\log\lambda_1 + q\log\lambda_2,$$

and logarithms are base 2.

The hunch is that c, the coefficient of growth, is maximized when $\lambda_1 = p$ and $\lambda_2 = q$. We prove this using the convexity of the log function:

$$c = 1 + p\log\lambda_1 + q\log\lambda_2$$

$$= 1 + p\log p + q\log q + p\log\frac{\lambda_1}{p} + q\log\frac{\lambda_2}{q}$$

$$\leq 1 + p\log p + q\log q + \log\left(p\frac{\lambda_1}{p} + q\frac{\lambda_2}{q}\right)$$

$$= 1 + p\log p + q\log q + \log(\lambda_1 + \lambda_2)$$

$$= 1 + p\log p + q\log q + \log 1$$

$$= 1 + p\log p + q\log q.$$

Hence, $c \leq 1 + p\log p + q\log q$, with equality if and only if $\lambda_1 = p$ and $\lambda_2 = q$.

With $p = 0.53$, the coefficient of growth is $c \doteq 0.00259841$. At this growth rate, it would take about 21 years of steadily playing the game to gain an expected return of \$1,000,000. Of course, you could win this amount in 20 days by betting the maximum every day, if you were very lucky; the chance is $(0.53)^{20}$, or about 3 out of 1,000,000.

Bonus: Channel Capacity

The value

$$c = 1 + p\log p + q\log q$$

in the Solution is significant in Information Theory. It is known as the *channel capacity* of a binary symmetric channel. The channel capacity measures the rate at which information can be reliably sent over a noisy channel.

In the *binary symmetric channel* (BSC), each binary symbol, 0 and 1, is sent accurately over the channel with probability p and inaccurately with probability $q = 1 - p$. See the picture below.

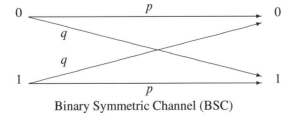

Binary Symmetric Channel (BSC)

The *capacity* $c(p)$ of the BSC is defined as

$$c(p) = 1 + p \log p + q \log q = 1 - H(p).$$

The graph of the capacity function $c(p)$ is shown below.

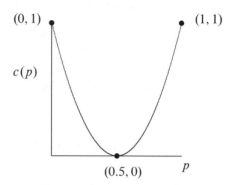

The capacity function $c(p)$.

Notice that $c(0.5) = 0$, which makes sense since a completely random channel cannot convey any information, and $c(0) = c(1) = 1$, which also makes sense since a channel that is 100% accurate or 100% inaccurate can convey information perfectly.

Shannon's Second Theorem states that information can be sent (using an error-correcting code) over a BSC with arbitrarily high accuracy at any rate below the channel capacity.

A Game for Noncommunicating Mathematicians

You are at the Joint Mathematics Meetings[4] when you are approached by an organizer who tells you that you may participate in a team game in which your team has the chance to win a valuable prize. The game is not a competitive one; your team will work together toward the goal of winning the prize. However, you may not meet with the other team members to discuss strategy or communicate with them in any way.

Specifically, you are told that there are ten team members (including yourself). Each player is given a fair coin. The players are not allowed to communicate with each other at any time. You will all gather in a room and, at a signal, each of you will open his/her hand to simultaneously reveal either the coin or nothing. Each player showing a coin will flip his/her coin. The team wins if at least one coin is flipped and all the flipped coins land heads. What strategy should each player follow in order to maximize the likelihood of the team winning? What is the probability of winning?

Solution

Let's solve the problem in general for a team of n players, where $n \geq 2$. Without the ban on communication, the team would choose one member to flip the coin and the team would win half of the time. However, the team cannot communicate, so it would appear

---

[4]The Joint Mathematics Meetings is an annual event in which members of several large mathematical organizations meet to give talks, discuss mathematics, and job hunt.

3.2 Probability

that there are only two possible strategies for each player: either flip the coin or don't flip it. But everyone not flipping the coin is bad because the team automatically loses, and everyone flipping the coin is bad because in that case the team wins with the very low probability $(1/2)^n$. With $n = 2$, the probability of winning if both players flip the coin is $1/4$. However, with a superior strategy a two-person team can do better than $1/4$. In fact, with best play, an n-person team can do better than if they were able to choose two members to flip their coins!

Each player should show his/her coin with probability p, where the value of p is to be determined ($0 \leq p \leq 1$). We obtain the probability of the team winning, $w_n(p)$, via a binomial expansion:

$$w_n(p) = \sum_{k=1}^{n} \left(\frac{1}{2}\right)^k \binom{n}{k} p^k (1-p)^{n-k}$$

$$= \left(\frac{p}{2} + 1 - p\right)^n - (1-p)^n$$

$$= \left(1 - \frac{p}{2}\right)^n - (1-p)^n. \quad (1)$$

Using calculus, we find that w_n is maximized at

$$p_n \equiv p = \frac{1 - \left(\frac{1}{2}\right)^{1/(n-1)}}{1 - \left(\frac{1}{2}\right)^{n/(n-1)}} \quad (2)$$

and the maximum winning probability is

$$w_n(p_n) = \left(2^{n/(n-1)} - 1\right)^{1-n}. \quad (3)$$

With two players, $p_2 = 2/3$ and $w_2(p_2) = 1/3$. You can use the coin to generate an event with the required probability: express p_n in base 2 and flip the coin until the number generated (heads = 1, tails = 0) deviates from p_n. The event is that the deviating bit is a 0. (This method was the subject of a 1990 Putnam Competition[5] problem. See [13].) You can handle the case $n = 2$ in a particularly simple way: flip the coin twice; if it lands tails both times, do over; the probability of one heads and one tails is $2/3$.

A calculation with l'Hôpital's rule (see Toolkit) shows that the winning probability w_n has the property

$$\lim_{n \to \infty} w_n(p_n) = \frac{1}{4}.$$

We can give an aha! explanation of this result. In (1), the second quantity is almost the square of the first quantity, so we can reason that the limit is of the form $x - x^2$, and anyone who can complete a square knows that the maximum value of $x - x^2$ is $1/4$.

It seems reasonable that the maximum probability of winning in (3) is a decreasing function of n. Let's define a real-variable version of the function:

$$f(x) = \left(2^{x/(x-1)} - 1\right)^{1-x}, \quad x > 1.$$

---

[5]The William Lowell Putnam Mathematical Competition is an annual mathematics contest open to students in the U.S.A. and Canada.

It's a tricky calculus problem to prove that $f(x)$ is decreasing. Surprisingly, we can give a short proof that $w_n(p_n)$ is decreasing via the solution to our original problem:

$$w_n(p_n) \geq w_n(p_{n+1}) = w_{n+1}(p_{n+1}).$$

The inequality holds by the definition of p_n. The equality, which you can prove by direct calculation combining (1) and (2), says that the success probabilities for n and $n+1$ are equal for the optimum probability p_{n+1}. This is tantamount to one team member "dropping out," but of course no one member can choose to do so.

We see from (2) that the optimum probability p_n decreases to 0 as n tends to infinity. What can we say about $n\, p_n$, the expected number of coins flipped? The limiting value follows from (2), using l'Hôpital's rule:

$$\lim_{n \to \infty} n\, p_n = \ln 4.$$

Furthermore, the expected number of coins flipped increases with n, that is,

$$n\, p_n < (n+1) p_{n+1}, \quad n \geq 2.$$

This inequality is equivalent to

$$n \cdot \frac{2^{n/(n-1)} - 2}{2^{n/(n-1)} - 1} < (n+1) \cdot \frac{2^{(n+1)/n} - 2}{2^{(n+1)/n} - 1},$$

which, by the formula for the sum of a geometric series, is equivalent to

$$\frac{1 + 2^{1/n} + 2^{2/n} + \cdots + 2^{n/n}}{n+1} < \frac{1 + 2^{1/(n-1)} + 2^{2/(n-1)} + \cdots + 2^{(n-1)/(n-1)}}{n}.$$

This last inequality follows from a simple proposition on convex functions applied to the function $\phi(x) = 2^x$.

PROPOSITION. *Let ϕ be a convex function defined on the interval $[0, 1]$. Then, for $n \geq 2$,*

$$\frac{1}{n+1} \sum_{i=0}^{n} \phi\left(\frac{i}{n}\right) \leq \frac{1}{n} \sum_{i=0}^{n-1} \phi\left(\frac{i}{n-1}\right).$$

If ϕ is strictly convex, then this inequality is strict.

This proposition is easy to prove using Jensen's inequality for convex functions (see p. 38).

Bonus: The Hat Problem

In the widely publicized Hat Problem (see, e.g., [3]), each contestant on a team tries to guess the color (blue or red) of a hat on his/her head while seeing only the hats of the other contestants. The guesses are made simultaneously and "passing" is an option. The team wins if at least one person guesses and no one guesses incorrectly. Before the guessing round, the participants are allowed to discuss strategy. With best play, an n-person team (where n is one less than a power of 2), can win with the surprisingly high probability $n/(n+1)$. The solution is related to the Fano Configuration that we will see in the problem "168 Elements."

3.3 Algebra

An Integer Matrix with Determinant 1

The Catalan numbers[6] C_n are given by the formula

$$C_n = \frac{1}{n+1}\binom{2n}{n}, \quad n \geq 0.$$

The Catalan sequence begins

$$1,\ 1,\ 2,\ 5,\ 14,\ 42,\ 132,\ 429,\ 1430,\ 5862,\ \ldots.$$

Catalan numbers occur in many combinatorial situations (see, for example, [19]). For instance, C_n is the number of paths in the x-y coordinate system that start at $(0, 0)$ and stop at (n, n), moving at each step either one unit right or one unit up and never crossing the line $y = x$. The figures below show the five different paths with $n = 3$. The lower-left point is $(0, 0)$ and the upper-right point is $(3, 3)$.

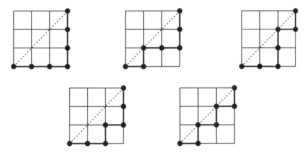

For any $n \geq 1$, form the $n \times n$ matrix whose (i, j)th entry is the Catalan number C_{i+j-2}. For example, with $n = 3$, we have the matrix

$$\begin{bmatrix} 1 & 1 & 2 \\ 1 & 2 & 5 \\ 2 & 5 & 14 \end{bmatrix}.$$

The determinant of this matrix is 1.

Prove that for every n the determinant of the matrix is 1.

Solution

Our solution combines the fact that the Catalan numbers count up-right paths (as above) with the permutation definition of the determinant. Recall that the determinant of an $n \times n$ matrix $A = [a(i, j)]$ is

$$\det A = \sum_\sigma \text{sgn}(\sigma) \prod_{i=1}^n a(i, \sigma(i)),$$

---

[6]The Catalan numbers are named after Eugène Charles Catalan (1814–1894), who studied them in connection with the Tower of Hanoi puzzle.

where σ ranges over all permutations of the set $\{1, 2, 3, \ldots, n\}$ and sgn(σ), the sign of σ, is 1 if σ is composed of an even number of transpositions and -1 if σ is composed of an odd number of transpositions.

We'll work out the computations for the $n = 3$ case in detail. The (i, j)th entry of our matrix is the Catalan number C_{i+j-2}, for $1 \le i, j \le 3$. Thus, it is the number of up-right paths starting at $(-(i-1), -(i-1))$, stopping at $(j-1, j-1)$, and not crossing the line $y = x$. Hence, a typical summand in the determinant is the number of systems of three paths from the points $(-(i-1), -(i-1))$ to the points $(\sigma(i)-1, \sigma(i)-1)$, for $1 \le i \le 3$, multiplied by sgn(σ).

The starting and stopping points are labeled in the diagram below.

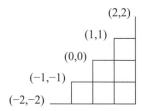

We claim that if any two paths in such a system have a point in common, then that system can be ignored in the determinant calculation. Suppose that a path starting at point a and stopping at point a' has a point, c, in common with a path starting at point b and stopping at point b'. Then switching the parts of the paths after c results in a path from a to b' and a path from b to a' which also have the point c in common. In the determinant calculation, the two corresponding summands are opposite in sign, as the transposition interchanging a' and b' switches the sign of the permutation. Hence, the contribution to the determinant from these systems of paths is 0. Therefore, we need only consider systems of three paths that have no points in common. This will make calculating the determinant very simple.

The next thing to observe is that only the identity permutation yields a system of three paths with no points in common. Consider the path that starts at $(-2, -2)$. If it stops at either of the points $(0, 0)$ or $(1, 1)$, then the path that stops at $(2, 2)$ will intersect it. Hence, we may assume that the path that starts at $(-2, -2)$ stops at $(2, 2)$. Similarly, we assume that the path that starts at $(-1, -1)$ stops at $(1, 1)$ and the path that starts at $(0, 0)$ stops at $(0, 0)$. This doesn't leave any wiggle room. The *only* allowable system of paths is the collection of nested right angles in the following diagram.

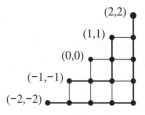

Since there is only one allowable system of paths (and it corresponds to the identity permutation), the determinant is 1. The same argument works for all n (the size of the matrix).

3.3 Algebra

Bonus: The Catalan Numbers Modulo 2 and 3

If we look at the Catalan numbers modulo 2, a pattern emerges:

n	0	1	2	3	4	5	6	7	8	...
C_n	1	1	2	5	14	42	132	429	1430	...
$C_n \bmod 2$	1	1	0	1	0	0	0	1	0	...

Based on the data, we guess that C_n is odd if and only if $n = 2^k - 1$, for some integer k. Let's prove this by induction. We will use the fact that the Catalan numbers satisfy the recurrence relation

$$C_n = \sum_{i=0}^{n-1} C_i C_{n-1-i}, \quad n \geq 1.$$

(This is easy to prove from the definition of Catalan numbers as counting up-right paths, as in the Problem).

The claim is true for $n = 0$, since $C_0 = 1$, an odd number, and $0 = 2^0 - 1$. Assume that the result holds for all Catalan numbers up to and including C_n. Now consider C_{n+1}. If $n + 1$ is even, then from the recurrence relation,

$$C_{n+1} = 2 \sum_{i=0}^{(n-1)/2} C_i C_{n-i},$$

which is even. If $n + 1$ is odd, then

$$C_{n+1} = 2 \sum_{i=0}^{(n-2)/2} C_i C_{n-i} + C_{n/2}^2,$$

which is odd if and only if $C_{n/2}$ is odd. Hence, C_{n+1} is odd if and only if $C_{n/2}$ is odd. By the induction hypothesis, this occurs if and only if $n/2 = 2^k - 1$, for some k, i.e., $n + 1 = 2^{k+1} - 1$. This completes the induction.

What about $C_n \bmod 3$? We will prove that the remainders modulo 3 of the Catalan sequence occur in groups of three identical terms:

$$C_{3k-1} \equiv C_{3k} \equiv C_{3k+1} \pmod{3}, \quad k \geq 1.$$

From the formula $C_n = \binom{2n}{n}/(n+1)$, we obtain a recurrence relation whereby each Catalan number can be computed from the next smaller number:

$$C_n = \frac{4n-2}{n+1} C_{n-1}, \quad n \geq 1.$$

(It's easy to prove this relation by evaluating the quotient C_n/C_{n-1} in terms of the formula and canceling lots of terms.) Multiplying by $n + 1$, we have

$$(n+1)C_n = (4n-2)C_{n-1},$$

and hence

$$(n+1)C_n \equiv (n+1)C_{n-1} \pmod{3}.$$

Letting $n = 3k$, we have $C_{3k} \equiv C_{3k-1} \pmod{3}$. Letting $n = 3k+1$, we obtain $C_{3k+1} \equiv C_{3k} \pmod{3}$. Therefore

$$C_{3k-1} \equiv C_{3k} \equiv C_{3k+1} \pmod{3}, \quad k \geq 1.$$

Much more can be discovered about the Catalan sequence modulo 3. Try to find more patterns yourself and see if you can prove them!

Only 1, −1, 0?

The four complex number solutions to the equation

$$z^4 = 1$$

are $1, -1, i$, and $-i$. The solution 1 satisfies the equation $z^1 = 1$, and the solution -1 satisfies the equation $z^2 = 1$. However, the solutions i and $-i$ do not satisfy any equation $z^n = 1$ with $0 < n < 4$. We call i and $-i$ *primitive 4th roots of unity*, and we create the monic polynomial of minimum degree that has these numbers as roots:

$$(z - i)(z + i) = z^2 + 1.$$

We call this polynomial, denoted $\Phi_4(z)$, the *cyclotomic polynomial* of order 4.

For $n \geq 1$, the primitive nth roots of unity are given by ζ^k, where $\zeta = e^{2\pi i/n}$ and $\gcd(k, n) = 1$. There are $\phi(n)$ primitive roots of order n (where ϕ is Euler's totient function). The *cyclotomic polynomial* of order n, denoted Φ_n, is the monic polynomial whose roots are the distinct primitive nth roots of unity:

$$\Phi_n(z) = \prod_{\substack{1 \leq k \leq n \\ \gcd(k,n)=1}} (z - \zeta^k).$$

The degree of $\Phi_n(z)$ is $\phi(n)$.

Prove that if n has at most two prime divisors, then the cyclotomic polynomial $\Phi_n(z)$ has no coefficients other than 1, −1, and 0. Find a value of n for which the cyclotomic polynomial $\Phi_n(z)$ has a coefficient other than 1, −1, and 0.

Solution

The key is a formula for computing cyclotomic polynomials:

$$z^n - 1 = \prod_{d \mid n} \Phi_d(z), \quad n \geq 1.$$

The formula works because every root of $z^n - 1$ is a primitive dth root of unity for *some unique* value of d, with $d \mid n$.

We will employ three consequences of this formula:
(1) For p prime, $\Phi_p(z) = z^{p-1} + \cdots + 1$.
(2) If n is odd, then $\Phi_{2n}(z) = \Phi_n(-z)$.

3.3 Algebra

(3) For $n = p_1^{e_1} \ldots p_k^{e_k}$ (canonical factorization), then

$$\Phi_n(z) = \Phi_{p_1 \ldots p_k}\left(z^{p_1^{e_1-1} \ldots p_k^{e_k-1}}\right).$$

Proof of (1): From the formula, we have

$$z^p - 1 = \Phi_1(z)\Phi_p(z) = (z-1)\Phi_p(z),$$

and hence

$$\Phi_p(z) = \frac{z^p - 1}{z - 1} = z^{p-1} + \cdots + 1.$$

Proof of (2): If ζ is a primitive nth root of unity, then $-\zeta$ is a primitive $2n$th root of unity. Furthermore, the number of primitive $2n$th roots of unity is the same as the number of nth roots of unity, as $\phi(2n) = \phi(2)\phi(n) = \phi(n)$. Hence

$$\Phi_{2n}(z) = \prod_{\gcd(k,n)=1} (z + \zeta^k) = (-1)^{\phi(n)} \prod_{\gcd(k,n)=1} (-z - \zeta^k) = \Phi_n(-z).$$

Proof of (3): We can "invert" our formula using Möbius inversion (see [5]) to obtain

$$\Phi_n(z) = \prod_{d|n} (z^d - 1)^{\mu(n/d)}, \quad n \geq 1,$$

where μ is the Möbius function, defined for nonnegative integers as follows:

$$\mu(n) = \begin{cases} 1 & \text{if } n = 1, \\ (-1)^k & \text{if } n \text{ is a product of } k \text{ distinct primes,} \\ 0 & \text{otherwise.} \end{cases}$$

If $\mu(n/d) \neq 0$, then n/d is square-free, so that $p_1^{e_1-1} \ldots p_k^{e_k-1} \mid d$. Thus we may restrict the product to divisors of the form $d' p_1^{e_1-1} \ldots p_k^{e_k-1}$ where $d' \mid p_1 \ldots p_k$, and so

$$\Phi_n(z) = \prod_{d' \mid p_1 \ldots p_k} (z^{p_1^{e_1-1} \ldots p_k^{e_k-1} d'} - 1)^{\mu(p_1 \ldots p_k/d')}$$

$$= \Phi_{p_1 \ldots p_k}(z^{p_1^{e_1-1} \ldots p_k^{e_k-1}} - 1).$$

We see from (1) and (3) that if n is divisible by only one odd prime, then the coefficients of $\Phi_n(z)$ are all 1. Now we will show that if n is divisible by only two distinct odd primes, then $\Phi_n(z)$ has coefficients only ± 1 and 0. We have

$$\Phi_{pq}(z) = \frac{(z^{pq} - 1)(z - 1)}{(z^p - 1)(z^q - 1)}.$$

We know that the rational function above is equal to a polynomial of degree $\phi(n) = (p-1)(q-1)$ with integer coefficients. We will show that the coefficients of this polynomial are all 1, −1, or 0. Our treatment is based on [14].

Since $\gcd(p, q) = 1$, there exist integers r and s such that

$$(p-1)(q-1) = rp + sq.$$

Furthermore, rewriting the above relation as

$$pq + 1 = (r+1)p + (s+1)q,$$

we see, since $pq + 1 > pq$, that we can take r and s to be nonnegative integers.

Let ζ be a primitive pqth root of unity. Then

$$\sum_{i=0}^{r}(\zeta^p)^i = -\sum_{i=r+1}^{q-1}(\zeta^p)^i, \quad \sum_{j=0}^{s}(\zeta^q)^j = -\sum_{j=s+1}^{p-1}(\zeta^q)^j,$$

and hence

$$\left(\sum_{i=0}^{r}\zeta^{ip}\right)\left(\sum_{j=0}^{s}\zeta^{jq}\right) - \zeta^{-pq}\left(\sum_{i=r+1}^{q-1}\zeta^{ip}\right)\left(\sum_{j=s+1}^{p-1}\zeta^{jq}\right) = 0.$$

Now consider the polynomial

$$f(z) = \left(\sum_{i=0}^{r}z^{ip}\right)\left(\sum_{j=0}^{s}z^{jq}\right) - z^{-pq}\left(\sum_{i=r+1}^{q-1}z^{ip}\right)\left(\sum_{j=s+1}^{p-1}z^{jq}\right).$$

Since an arbitrary primitive pqth root of unity, ζ, is a root of this polynomial, and the degree of the polynomial is $(p-1)(q-1) = \phi(pq)$, it follows that $f(z) = \Phi_{pq}(z)$. By inspection, each coefficient of the polynomial is $1, -1$, or 0.

It follows from our work above and formulas (1), (2), and (3), that if a cyclotomic polynomial has coefficients other than ± 1 and 0, then its order is divisible by three distinct prime numbers. The smallest number with three distinct odd prime factors is 105, and indeed the cyclotomic polynomial $\Phi_{105}(z)$ has the terms $-2z^7$ and $-2z^{41}$.

Bonus: Flat Cyclotomic Polynomials

The condition that the order n of a cyclotomic polynomial is divisible by three distinct primes is necessary for the polynomial to have coefficients other than ± 1 and 0, but it is not a sufficient condition. In fact, in 2006 Gennady Bachman [1] proved the existence of an infinite family of cyclotomic polynomials of order pqr (where p, q, r are distinct odd primes) with coefficients only $-1, 0,$ and 1. Such polynomials are called "flat polynomials." The smallest example has order $3 \cdot 7 \cdot 11 = 232$.

168 Elements

Let G be the group given by the presentation

$$G = \langle s, t : s^7 = t^2 = (st)^3 = (s^4t)^4 = 1 \rangle.$$

Show that G has 168 elements.

Note. The elements of G are all possible finite strings of s's and t's. For example, G contains the string

$$sssttt stst st.$$

3.3 Algebra

Multiplication occurs by concatenating strings. If $x \in G$, then we denote xx by x^2, xxx by x^3, etc. We denote x^0 by 1.

Because of the equations $t^2 = 1$ and $(st)^3 = 1$, our example string $ssstttststst$ reduces to the string

$$s^3 t.$$

In this problem, we want to show that every string of s's and t's reduces to one of 168 different strings.

Solution

The "aha!" realization consists of two parts. The first is a criterion for reducing strings. The second is a method of showing that the reduced strings are distinct.

We will show that every element of G is equal to a string of one of the following types:
(A) Strings of the form s^a. There are 7 such strings.
(B) Strings of the form $s^a t s^b$. There are $7 \times 7 = 49$ such strings.
(C) Strings of the form $s^a t s^2 t s^b$ or $s^a t s^3 t s^b$. There are $7 \times 7 \times 2 = 98$ such strings.
(D) Strings of the form $t s^2 t s^4 t s^b$ or $t s^4 t s^2 t s^b$. There are $7 \times 2 = 14$ such strings.

The total number of such strings is 168.

At this point we do not claim that the above strings are all distinct. In fact they are, but this follows from the existence of a representation of G by the matrix group $GL(3, 2)$, which has 168 elements.

So our task is to prove that every element of G is equal to one of the above strings. An element of G may be written as a string in many ways. Let's consider only strings with a minimum number of occurrences of t. Note that because of the equation $t^2 = 1$, an element of G consists of alternating powers of s (where the exponents are between 1 and 6) and terms of t.

We can say that certain substrings of an element do not occur. For example, the substring $ts^6 t$ does not occur, since the equation $(st)^3 = 1$ implies that $ts^6 t = sts$, and the expression on the right has only one t (versus the two t's on the left).

The method of proof is to show that many substrings do not occur, leaving as the only possibilities for strings those on our list (A)–(D).

Let's identify two equations from the presentation, with several variations:
(1) $(st)^3 = 1$
(1a) $ststst = 1$
(1b) $ts^6 t = sts$
(1c) $tst = s^6 t s^6$
(2) $(s^4 t)^4 = 1$
(2a) $s^4 t s^4 t s^4 t s^4 t = 1$.
(2b) $ts^4 t s^4 t = s^3 t s^3$
(2c) $ts^3 t s^3 t = s^4 t s^4$
(2d) $ts^4 t = s^3 t s^3 t s^3$.

Now we go about describing the strings that do not occur. From (1b), the string $ts^6 t$ does not occur, and from (1c), the string tst does not occur. We abbreviate this by saying that 6 and 1 do not occur, where these numbers refer to the exponents of s. To be precise, when we say that a number or a string of numbers does not occur, we mean that the string consisting of the corresponding powers of s, separated by t's, and beginning and ending with a t, does not occur. From (2d), we see that we can change a 4 to a 3 (and *vice versa*).

From (2b), the string 44 does not occur, and from (2c), the string 33 does not occur.

From (1c), we obtain

(3) $ts^2t = tsttst = s^6ts^5ts^6$.

From (3), we can change a 2 to a 5 (and *vice versa*).

From (1c), we obtain

(4) $ts^2ts^2t = s^6ts^5ts^6s^2t = s^6ts^5tst = s^6ts^5s^6ts^6 = s^6ts^4ts^6$.

Thus, 22 does not occur.

From (1c) and (2d), we obtain

(5) $ts^5t = tstts^4t = s^6ts^6s^3ts^3ts^3ts^3 = s^6ts^2ts^3ts^3$

and, alternatively,

(6) $ts^5t = ts^4ttst = s^3ts^3ts^3s^6ts^6 = s^3ts^3ts^2ts^6$.

Now we see from (5) that 23 does not occur, and from (6) that 32 does not occur.

From (2d), we obtain $ts^3ts^4t = ts^3(s^3ts^3ts^3) = ts^6ts^3ts^3$, which reduces since the right side contains a 6. Hence, 34 does not occur.

From (3), we obtain $ts^2(ts^5t) = ts^3s^6(ts^5t)s^6s = ts^3(s^6ts^5ts^6)s = ts^3ts^2ts$, which reduces since the right side contains a 32. Hence, 25 does not occur.

By symmetry, 43 does not occur. (The symmetry comes from the symmetry in the exponents of s in the formulas (3) and (2d).)

From (2d) and (1c), we obtain $ts^4ts^5t = s^3ts^3tst = s^3t\dot{s}^3s^6ts^6$. Hence 45 does not occur.

By symmetry, 52 does not occur.

From (3), we obtain $ts^5ts^5t = ss^6ts^5ts^6s^6t = sts^2ts^6t$, which reduces since the right side contains a 6. Hence, 55 does not occur.

To summarize, the strings that we have shown do not occur are 1, 6, 22, 23, 25, 32, 33, 34, 43, 44, 45, 52, 54, and 55. The only pairs left are 24, 35, 42, and 53.

From (2d), we obtain

$$ts^3ts^5ts^3t = s^4(s^3ts^3ts^3)s^6(s^3ts^3ts^3) = s^4(ts^4t)s^6(ts^4t),$$

which reduces since it contains a 6.

Hence, 353 does not occur. Similarly, 535 does not occur.

From (3), we obtain

$$(ts^2t)s^4(ts^2t) = (s^6ts^5ts^6)s^4(s^6ts^5ts^6) = s^6ts^5ts^2ts^5ts^6,$$

which reduces since it contains a 25.

Hence, 242 does not occur. Similarly, 424 does not occur.

We have established that the number of words is finite, since no word with four t's can occur (we have excluded all triples of exponents of s). Thus, the non-excluded words have zero, one, two, or three t's. The cases of zero and one t are, respectively, (A) and (B). The case of two t's looks like $s^ats^cts^b$, where $c = 2, 3, 4,$ or 5. However, 2 can become 5 (by (3)) and 3 can become 4 (by (2d)), so in fact we have $c = 2, 3$. These strings comprise (C). The case of three t's looks like $s^ats^2ts^4ts^b$ or $s^ats^4ts^2ts^b$ (the only pairs are 24

3.3 Algebra

and 42, as 35 can become 42 and 53 can become 24). In fact, we can take $a = 0$ via the transformation

$$s^a t s^2 t s^4 s^b = s^a (s^6 t s^5 t s^6) s^4 t s^b \quad \text{(by (3))}$$
$$= s^{a-1} t s^5 t s^3 t s^b,$$

followed by changing 53 to 24 (by (2c)):

$$s^{a-1} t s^5 t s^3 t s^b = s^{a-1} t s^2 t (t s^3 t s^3 t) s^b$$
$$= s^{a-1} t s^2 t s^4 t s^4 s^b$$
$$= s^{a-1} t s^2 t s^4 t s^{b+4}.$$

Performing this transformation a times results in a string with $a = 0$. The case 42 is dealt with similarly. These cases are thus covered by (D).

Now we know that G consists of at most 168 elements, as evidenced by our set of reduced strings. We now need to show that these strings are distinct. This we do by furnishing a representation of G with 168 elements. The group is $GL(3, 2)$, the group of 3×3 invertible matrices over the field $\{0, 1\}$.

The *general linear group* $GL(n, q)$ is the multiplicative group of invertible $n \times n$ matrices with coefficients in the field of order $q = p^k$, where p is a prime. Let's determine the order of $GL(n, q)$. There are $q^n - 1$ choices for the first row of an invertible $n \times n$ matrix (the all 0 row is excluded). Given the first row, the second row can be any of the q^n possible n-tuples except the q scalar multiples of the first row. Hence, there are $q^n - q$ choices for the second row. Similarly, the third row can be any of the q^n n-tuples except linear combinations of the first two rows. There are $q^n - q^2$ choices. Continuing in this manner, we find that the number of invertible matrices is

$$|GL(n, q)| = (q^n - 1)(q^n - q)(q^n - q^2) \ldots (q^n - q^{n-1}).$$

In particular

$$|GL(3, 2)| = (2^3 - 1)(2^3 - 2)(2^3 - 2^2)$$
$$= 7 \cdot 6 \cdot 4$$
$$= 168.$$

For example, $GL(2, 2)$ has six elements and is in fact isomorphic to S_3. The group Aut FC (to be defined) is isomorphic to $GL(3, 2)$.

The Fano Configuration is named after the geometer Gino Fano (1871–1952).[7] The Fano Configuration is the prototypical example of many types of combinatorial structures, such as projective planes, block designs, and difference sets. Let's look at the configuration and observe some of its properties.

The Fano Configuration (FC) is shown below.

---

[7] In Fano's geometry, what we call the Fano Configuration was specifically excluded. This situation reminds me of that of chess master Pedro Damiano (1480–1544), who analyzed a particular chess opening, found it wanting, and recommended against playing it, only to have it named after him (Damiano's Defense). Such are the vagaries of naming conventions.

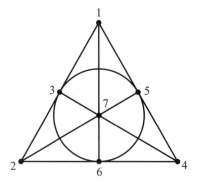

The points of FC are labeled 1, 2, 3, 4, 5, 6, and 7. The lines are just unordered triples of points (e.g., $\{1, 2, 3\}$); they have no Euclidean meaning.

We observe that FC has the following properties:

1. There are seven points.
2. There are seven lines.
3. Every line contains three points.
4. Every point lies on three lines.
5. Every two points lie on exactly one line.
6. Every two lines intersect in exactly one point.

The above properties occur in pairs called *duals*. If the words 'point' and 'line' are interchanged, each property is transformed into its dual property.

We next determine the automorphism group of FC, denoted Aut FC. This automorphism group is the group of permutations of the vertices of FC which preserve collinearity. For example, if we rotate FC clockwise by one-third of a circle, collinearity is preserved. This particular permutation is written in cycle notation as $(142)(356)(7)$.

We begin by calculating the order of Aut FC. It is evident from the picture of FC that all seven vertices are equivalent in terms of collinearity. Therefore, vertex 1 may be mapped by an automorphism to any of the seven vertices. Suppose that 1 is mapped to $1'$. Vertex 2 may be mapped to any of the remaining six vertices. Suppose that 2 is mapped to $2'$. In order to preserve collinearity, vertex 3 must be mapped to the unique point collinear with $1'$ and $2'$. Call this point $3'$. Vertex 4 is not on line 123, so its image $4'$ can be any of the remaining four vertices. Finally, the images of the other points are all determined by collinearity: 5 is collinear with 1 and 4; 6 is collinear with 2 and 4; and 7 is collinear with 3 and 4. Hence, there are $7 \cdot 6 \cdot 4 = 168$ automorphisms.

Now we know that Aut FC is a group of order 168. In fact, Aut FC is isomorphic to a group of matrices over a finite field.

Let F be the field of two elements $\{0, 1\}$, and let F^n be the vector space of n-tuples of elements of F. A vector $v \in F^n$ is represented by a string of length n over F, e.g., $v = 010101 \in F^6$. The cardinality of F^n is 2^n.

We will now show that Aut FC is isomorphic to $GL(3, 2)$. We label the vertices of FC with the seven nonzero vectors in F^3, as shown below. This labeling is derived from the labeling in the previous picture by assigning to each vertex i the vector which represents the number i in binary. The vectors have been chosen so that v_1, v_2, v_3 are collinear if

3.3 Algebra

and only if $v_1 + v_2 + v_3 = 0$ (in F^3). The matrix group GL(3, 2) acts on the vectors of FC by the rule: $v \mapsto vM$. It is easy to check that this action preserves collinearity: $v_1 + v_2 + v_3 = 0 \Leftrightarrow (v_1 + v_2 + v_3)M = 0 \cdot M \Leftrightarrow v_1 M + v_2 M + v_3 M = 0$. Therefore, Aut FC is isomorphic to the group of 3×3 invertible matrices over F.

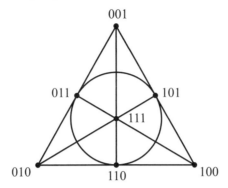

The *special linear group* SL(n, q) is the normal subgroup of GL(n, q) consisting of $n \times n$ invertible matrices with entries from the field of q elements and determinant 1. The quotient group is given by

$$\text{GL}(n,q)/\text{SL}(n,q) \simeq F \setminus \{0\},$$

where F is the field of q elements, and hence

$$|\text{SL}(n,q)| = \frac{(q^n - 1)(q^n - q)(q^n - q^2)\ldots(q^n - q^{n-1})}{q - 1}.$$

For all $n \geq 1$, we have GL($n, 2$) $\simeq$ SL($n, 2$).

In fact (see [18]), SL($n, 2$) is a simple group (a group with no nontrivial normal subgroups), for all $n \geq 2$. The simple groups are crucial to the study of algebra. FC is a geometric model of SL(3, 2) ($\simeq$ GL(3, 2)).

Let's investigate possible generators for Aut FC. By direct calculation we see that the rows of M are the images of the binary representations of 1, 2, and 4. Let

$$S = \begin{bmatrix} 0 & 1 & 1 \\ 1 & 0 & 0 \\ 1 & 0 & 1 \end{bmatrix} \text{ and } T = \begin{bmatrix} 1 & 0 & 0 \\ 0 & 0 & 1 \\ 0 & 1 & 0 \end{bmatrix}.$$

We observe that T yields a flip of FC around the 374 axis while S yields the 7-cycle (1 5 6 7 2 4 6). We will show that all 168 matrices in the automorphism group of FC are generated by combinations of S and T.

The "visible automorphisms" of FC are the symmetries of the equilateral triangular given by the elements of the symmetric group S_3. These symmetries are combinations of S and T, for T is a reflection of the triangle (i.e., a transposition of two of its vertices) and the element $(S^2TS^2)^2T$ is a rotation of the triangle by one-third of a circle; all symmetries of the triangle (permutation matrices) are combinations of a reflection and a rotation. Furthermore, the element STS^2T is a transvection (the identity matrix with an extra 1 in an off-diagonal position). From the transvection, we can produce all transvections via conjugation by permutations. Using elementary row operations, all invertible matrices can be formed from permutation matrices and transvections. Therefore, all invertible matrices are combinations of S and T.

Bonus: A Tiling of the Hyperbolic Plane

The element ST has order 3 (but isn't a rotation). Letting $u = (st)^{-1}$, we can say that the group of order 168 is a homomorphic image of the "triangle group" generated by s, t, and u:

$$\langle s, t, u : s^7 = t^2 = u^3 = stu = 1 \rangle.$$

This is the group of symmetries of a tiling of the hyperbolic plane with triangles with angles $\pi/2$, $\pi/3$ and $\pi/7$. This tiling is shown below.

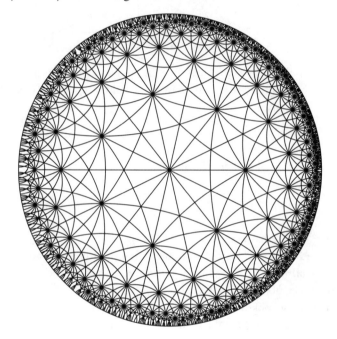

3.4 Number Theory

Odd Binomial Coefficients

Let k and m be integers such that $0 \leq k \leq 2^m - 1$. Prove that the binomial coefficient $\binom{2^m-1}{k}$ is odd.

Solution

A much more general result is true! We will prove that the binomial coefficient $\binom{n}{k}$ is odd if and only if the binary representation of n contains a 1 in every position where the binary representation of k contains a 1. The result called for follows instantly.

3.4 Number Theory

Suppose that the base-2 representation of n is

$$n = \sum_{i=0}^{l} n_i 2^i.$$

By de Polignac's formula (see Bonus) the exact power of 2 that divides $n!$ is

$$\sum_{i=1}^{l} n_i 2^{i-1} + \sum_{i=2}^{l} n_i 2^{i-2} + \cdots + \sum_{i=l}^{l} n_i 2^{i-l} = \sum_{i=1}^{l} n_i \left(2^{i-1} + 2^{i-2} + \cdots + 1\right)$$

$$= \sum_{i=0}^{l} n_i (2^i - 1)$$

$$= \sum_{i=0}^{l} n_i 2^i - \sum_{i=0}^{l} n_i$$

$$= n - d_2(n),$$

where $d_2(n)$ is the number of 1's in the binary representation of n.

Next, we will show that the exact power of 2 that divides the binomial coefficient $\binom{n}{k}$ is equal to the number of 'carries' when k and $n-k$ are added in binary.[8] By the preceding observation, this power is

$$n - d(n) - (k - d(k) + n - k - d(n-k)) = d(k) + d(n-k) - d(n).$$

Denote the ith bit of a binary number x by x_i. Let $c(i) = 1$ if there is a carry in the ith bit when k and $n-k$ are added (in binary), and $c(i) = 0$ if there is no carry, for $0 \leq i \leq l$. Also, define $c(-1) = 0$. We see that $n_i = k_i + (n-k)_i + c(i-1) - 2c(i)$. Hence, the power to which 2 divides $\binom{n}{k}$ is

$$\sum_{i=0}^{l} [k_i + (n-k)_i - n_i] = \sum_{i=0}^{l} (2c(i) - c(i-1)) = \sum_{i=0}^{l} c(i).$$

Thus, we conclude that $\binom{n}{k}$ is odd if and only if there are no carries when k and $n-k$ are added in binary. This is the case precisely when the binary representation of n contains a 1 in every position where the binary representation of k contains a 1.

Bonus: De Polignac's Formula.[9]

Let p be a prime. Then the exact power of p that divides $n!$ is

$$\sum_{j=1}^{\infty} \left\lfloor \frac{n}{p^j} \right\rfloor.$$

This is known as de Polignac's formula.

---

[8] This result was discovered by Ernst Eduard Kummer (1810–1893), who made contributions to algebraic geometry and number theory.

[9] The formula is credited to Alphonse de Polignac (1817-1890), who is best known for making the Twin Prime Conjecture: There are infinitely pairs of primes whose difference is 2, e.g., **17** and **19**. This conjecture is unproved.

Note. The sum is actually finite, since $p^j > n$ for j sufficiently large.

For example, the exact power of 3 that divides 100! is

$$\left\lfloor \frac{100}{3} \right\rfloor + \left\lfloor \frac{100}{9} \right\rfloor + \left\lfloor \frac{100}{27} \right\rfloor + \left\lfloor \frac{100}{81} \right\rfloor = 33 + 11 + 3 + 1 = 48.$$

Thus, 100! is divisible by 3^{48} but not by 3^{49}.

To prove the formula, we observe that there are exactly $\lfloor n/p \rfloor$ multiples of p less than or equal to n, exactly $\lfloor n/p^2 \rfloor$ multiples of p^2 less than or equal to n, and in general, exactly $\lfloor n/p^j \rfloor$ multiples of p^j less than or equal to n. The formula follows immediately.

Fibonacci Factors

Show that no Fibonacci number f_n with n odd has a prime factor of the form $4k + 3$.

Solution

We check the claim by finding the prime factorizations of the first ten Fibonacci numbers with odd subscripts:

f_1	f_3	f_5	f_7	f_9	f_{11}	f_{13}	f_{15}	f_{17}	f_{19}
1	2	5	13	34	89	233	610	1597	4181
1	2	5	13	$2 \cdot 17$	89	233	$2 \cdot 5 \cdot 61$	1597	$37 \cdot 113$

By inspection, no prime factor above is of the form $4k + 3$.

So we have data to support the claim, but how do we prove the claim? A solution ties together some tenuous clues. What distinguishing features of primes of the form $4k + 3$ do you know? One such feature is that for no prime p of the form $4k + 3$ is the congruence

$$x^2 \equiv -1 \pmod{p}$$

solvable for an integer x. On the other hand, the same congruence is always solvable for primes of the form $4k + 1$. For instance, $4^2 \equiv -1 \pmod{17}$.[10]

So now we see that the proof of the claim may have something to do with squares modulo p. Can we think of a formula involving squares of Fibonacci numbers?

An identity comes to the rescue! Just the right sort of formula is supplied by Cassini's identity:[11]

$$f_n^2 - f_{n+1} f_{n-1} = (-1)^{n+1}, \quad n \geq 1.$$

It is easy to prove Cassini's formula by induction, but we give an aha! proof using determinants. The Fibonacci numbers are conveniently generated by the matrix equation

$$\begin{bmatrix} f_{n+1} & f_n \\ f_n & f_{n-1} \end{bmatrix} = \begin{bmatrix} 1 & 1 \\ 1 & 0 \end{bmatrix}^n, \quad n \geq 1.$$

---

[10] The reason that the existence of a solution to this congruence depends on the mod 4 status of p has to do with the fact that the $p - 1$ nonzero residue classes modulo p form a cyclic group.

[11] This identity was discovered by the astronomer Giovanni Domenico Cassini (1625–1712).

3.4 Number Theory

(This is easily proved by induction.) Take the determinant of both sides, using the rule that the determinant of a product is the product of the determinants:

$$f_{n+1} f_{n-1} - f_n^2 = \begin{vmatrix} f_{n+1} & f_n \\ f_n & f_{n-1} \end{vmatrix} = \begin{vmatrix} 1 & 1 \\ 1 & 0 \end{vmatrix}^n = (-1)^n.$$

This yields Cassini's identity!

Armed with Cassini's identity, we can easily prove the claim. If $p \mid f_{n-1}$ and n is even, then

$$f_n^2 \equiv -1 \pmod{p}.$$

But this is impossible if $p = 4k + 3$, as we have stated.

By the way, Wayne McDaniel has shown that all but eleven Fibonacci numbers have a prime factor of the form $4k + 1$.

Bonus: Repeated Numbers in Pascal's Triangle

We use Cassini's identity to prove a gem concerning Pascal's triangle.

The number 2 occurs in the second row of Pascal's triangle. It is evident that 2 will not occur elsewhere, for the smallest number in row n (barring the 1's at the ends) is n. How often does, say, the number 10 appear? The fifth row of Pascal's triangle contains two 10's and the tenth row contains two 10's. A search of the first ten rows shows that 10 occurs exactly four times. Similarly, any number (greater than 3) that appears twice as the two middle terms in a row appears at least four times altogether. However, it is difficult to find numbers that appear more than four times. Would you be amazed to learn that there are infinitely many numbers that occur at least six times in Pascal's triangle?

Consider solutions to

$$t = \binom{n}{m-1} = \binom{n-1}{m}, \quad 3 \leq m < (n-1)/2.$$

When there exists such a solution, the number t appears at least six times in Pascal's triangle: twice in row t, twice in row n (in columns $m-1$ and $n-m+1$), and twice in row $n-1$ (in columns m and $n-1-m$).

Solutions are given by

$$m = f_{2k-1} f_{2k}, \quad n = f_{2k} f_{2k+1}, \quad k \geq 2.$$

We prove this by checking that the following equations are equivalent:

$$\binom{n}{m-1} = \binom{n-1}{m},$$

$$\frac{n!}{(m-1)!(n-m+1)!} = \frac{(n-1)!}{m!(n-m-1)!},$$

$$mn = (n - m + 1)(n - m),$$

$$f_{2k-1} f_{2k} f_{2k} f_{2k+1} = (f_{2k} f_{2k+1} - f_{2k-1} f_{2k} + 1)(f_{2k} f_{2k+1} - f_{2k-1} f_{2k})$$
$$= (f_{2k}(f_{2k+1} - f_{2k-1}) + 1)(f_{2k}(f_{2k+1} - f_{2k-1}))$$
$$= (f_{2k}^2 + 1) f_{2k}^2,$$

$$f_{2k-1} f_{2k+1} = f_{2k}^2 + 1.$$

The last relation is Cassini's identity (with n replaced by $2k$). For $k = 2$, we obtain the value $t = 3003$. Hence, 3003 appears at least six times in Pascal's triangle.

In fact, 3003 appears eight times in Pascal's triangle, in the following representations:

$$\binom{14}{6} = \binom{14}{8} = \binom{15}{5} = \binom{15}{10} = \binom{78}{2} = \binom{78}{76} = \binom{3003}{1} = \binom{3003}{3002}.$$

No other numbers are known to appear as many as eight times in Pascal's triangle, and the maximum number of appearances of a number in Pascal's triangle is also unknown.

Exact Covering Systems

A *covering system* is a collection of congruences of the form $x \equiv a_i \pmod{m_i}$, for $1 \leq i \leq k$, where the m_i are integers greater than 1, such that every integer x satisfies at least one of the congruences. For example, the congruences $x \equiv 0 \pmod{2}$, $x \equiv 0 \pmod{3}$, $x \equiv 1 \pmod{4}$, $x \equiv 1 \pmod{6}$, and $x \equiv 11 \pmod{12}$ constitute a covering system. It's easy to verify this. Just write the integers 0 through 11 and cross out those covered by each congruence. All the numbers will be crossed out.

A covering system in which each integer x satisfies exactly one of the congruences is called an *exact covering system*. The congruences $x \equiv 0 \pmod{2}$ and $x \equiv 1 \pmod{2}$ are an exact covering system (but a boring one).

Prove that there does *not* exist an exact covering system with distinct moduli.

Solution

Here is a beautiful and surprising proof using generating functions. Let $m_1, m_2, \ldots, m_k$ be the distinct moduli of the congruences in an exact covering system. For $1 \leq i \leq k$, the positive integers that satisfy the congruence $x \equiv a_i \pmod{m_i}$ are represented by the generating function

$$x^{a_i} + x^{a_i + m_i} + x^{a_i + 2m_i} + x^{a_i + 3m_i} + \cdots = \frac{x^{a_i}}{1 - x^{m_i}}.$$

Therefore,

$$\sum_{i=1}^{k} \frac{x^{a_i}}{1 - x^{m_i}} = \frac{1}{1 - x},$$

since $1/(1 - x)$ is the generating function for the set of all positive integers.

Now, the right side is a rational function with only one discontinuity (called a pole), at $x = 1$. Let m be the maximum of the m_i (which are distinct). On the left side, the

3.4 Number Theory

term $x^{am}/(1-x^m)$ has a discontinuity at $x = e^{2\pi i/m}$, and no other discontinuity cancels this out. This is a contradiction. Therefore, there is no exact covering system with distinct moduli.

Bonus: Never a Prime

Is it true that given any odd integer k, there exists an integer n such that $k + 2^n$ is a prime number? For $k = 5$ we may take $n = 1$, since $5 + 2^1 = 7$, a prime. For $k = 7$ we may take $n = 2$, since $7 + 2^2 = 11$, a prime. Does this always work? We will follow a method of Paul Erdős, using a covering system of congruences, to find an odd integer k such that $k + 2^n$ is prime for *no* n.

To get a feeling for what is going on, take $k = 83$. Now, $83 + 2^1 = 85$, which is divisible by 5. Hence, $83 + 2^n$ is divisible by 5 if $2^n \equiv 2 \pmod{5}$. The powers of 2 modulo 5 form the cycle $\{2, 4, 3, 1\}$. It follows that $83 + 2^n$ is composite (divisible by 5) for $n \equiv 1 \pmod{4}$. Also, $83 + 2^2 = 87$, which is divisible by 3. Hence, $83 + 2^n$ is divisible by 3 if $2^n \equiv 1 \pmod{3}$. The powers of 2 modulo 3 form the cycle $\{2, 1\}$. It follows that $k + 2^n$ is composite (divisible by 3) for $n \equiv 0 \pmod{2}$. We have so far ruled out two infinite arithmetic progressions as choices for k, namely, all solutions to $n \equiv 1 \pmod{4}$ and $n \equiv 0 \pmod{2}$. The smallest positive integer not ruled out is 7, and $83 + 2^7 = 211$, a prime.

The idea of Erdős' method is to choose sufficiently many primes to rule out enough arithmetic progressions to eliminate all possible values of n. We see from the example $k = 83$ that we should study cycles of powers of 2 modulo various primes. This is easy to do by hand or on a computer, but we probably want to use a computer to complete our study since one of the key primes is 241, yielding a cycle of 24 powers of 2. The following table shows our selection of primes and the number of powers of 2 modulo these primes.

prime	length of cycle of powers of 2
3	2
5	4
7	3
13	12
17	8
241	24

We make a covering system with the lengths as moduli, as follows:

$$x \equiv 1 \pmod{2}$$

$$x \equiv 0 \pmod{4}$$

$$x \equiv 0 \pmod{3}$$

$$x \equiv 2 \pmod{12}$$

$$x \equiv 2 \pmod{8}$$

$$x \equiv 22 \pmod{24}.$$

To find an odd integer k such that $k + 2^n$ is never prime, we look for a solution to the system of congruences

$$k \equiv 1 \pmod{2}$$
$$k \equiv -2^1 \pmod{3}$$
$$k \equiv -2^0 \pmod{5}$$
$$k \equiv -2^0 \pmod{7}$$
$$k \equiv -2^2 \pmod{13}$$
$$k \equiv -2^2 \pmod{17}$$
$$k \equiv -2^{22} \pmod{241}.$$

(The purpose of the first congruence is to ensure that k is odd.) This will furnish an arithmetic progression of integers k that satisfy the condition. For suppose, say, that $n \equiv 0 \pmod{4}$. Then $2^n \equiv 1 \pmod{5}$ and since $k \equiv -1 \pmod{5}$, it follows that $k + 2^n$ is divisible by 5 (and hence composite).

We find a simultaneous solution to the above congruences using the Chinese remainder theorem. Let $m = 2 \cdot 3 \cdot 5 \cdot 7 \cdot 13 \cdot 17 \cdot 241 = 11184810$. Next we solve the following congruences:

$$(m/2)k_1 \equiv 1 \pmod{2}$$
$$(m/3)k_2 \equiv -2^1 \pmod{3}$$
$$(m/5)k_3 \equiv -2^0 \pmod{5}$$
$$(m/7)k_4 \equiv -2^0 \pmod{7}$$
$$(m/13)k_5 \equiv -2^2 \pmod{13}$$
$$(m/17)k_6 \equiv -2^2 \pmod{17}$$
$$(m/241)k_7 \equiv -2^{22} \pmod{241}.$$

We find the solutions $k_1 = 1, k_2 = 2, k_3 = 2, k_4 = 2, k_5 = 12, k_6 = 1$, and $k_7 = 210$. Therefore, the solution, modulo m, is

$$k \equiv \left(\frac{m}{2}\right)1 + \left(\frac{m}{3}\right)2 + \left(\frac{m}{5}\right)2 + \left(\frac{m}{7}\right)2 + \left(\frac{m}{13}\right)12 + \left(\frac{m}{17}\right)1 + \left(\frac{m}{241}\right)210 = 41446999.$$

Let's take k to be the smallest positive integer in this congruence class, i.e.,

$$k = 41446999 - 3 \cdot 11184810 = 7892569.$$

So $7892569 + 2^n$ is prime for no n. Can you find a smaller value of k?

A Fibonacci Number Producing Polynomial

In 1975 James P. Jones wrote an article titled "Diophantine representation of the Fibonacci numbers" for the *The Fibonacci Quarterly* [12]. Paul Hartung, in his review of that article, said "The author throws a real bombshell here." The bombshell was the following: The set

3.4 Number Theory

of positive values of the polynomial

$$2y - x^4y - 2x^3y^2 + x^2y^3 + 2xy^4 - y^5,$$

where x and y are positive integers, is precisely the set of (positive) Fibonacci numbers. Furthermore, the polynomial represents each Fibonacci number exactly once. Can you explain why this is true?

Solution

We begin by writing the polynomial in the useful form

$$y[2 - (y^2 - xy - x^2)^2].$$

Now it's easy to see that the polynomial can have a positive value if and only if $y^2 - xy - x^2 = 0$ or ± 1. But the first case is impossible, since $y^2 - xy - x^2 = 0$ implies that $4y^2 - 4xy - 4x^2 = 0$, and hence $(2y - x)^2 = 5x^2$, but 5 is not the square of an integer. Therefore, we must have

$$y^2 - xy - x^2 = \pm 1.$$

Does this equation remind you of a famous identity?

Cassini's identity (recall the solution to "Fibonacci Factors") is

$$f_n^2 - f_{n-1} f_{n+1} = \begin{cases} 1 & n \text{ odd,} \\ -1 & n \text{ even.} \end{cases}$$

Let's rewrite the relation as

$$y^2 - x(y + x) = \pm 1.$$

Now we see the resemblance to Cassini's identity.

By Cassini's identity, the ordered pairs $(x, y) = (f_{n-1}, f_n)$, for $n \geq 1$, satisfy the equation. In fact, these are the only solutions.

The main idea is that if (x, y) satisfies the equation, then $(y - x, x)$ satisfies the same equation, since

$$x^2 - (y - x)x - (y - x)^2 = x^2 - (y - x)y = x^2 + xy - y^2 = \mp 1.$$

Continuing the descent, we end up with $(x, y) = (1, 2)$. Ascending from this pair, the transformation $(x, y) \to (y, x + y)$ yields the sequence $\{(f_{n-1}, f_n)\}$, with n even if the $+1$ occurs and n odd if the -1 occurs.

Bonus: Hilbert's Tenth Problem

Consider the equation

$$2x + 3y = 100.$$

We can see that the equation has integer solutions, e.g., $(x, y) = (35, 10)$. By contrast, the equation

$$x^2 + y^2 = 103$$

has no integer solutions, as we can see by checking all values of x, y such that $0 \leq x, y \leq 10$. Another way to see that this equation has no solutions is to recognize that squares always are always congruence to 0 or 1 modulo 4. Since the left side is the sum of two squares, it is congruent to 0, 1, or 2 modulo 4, while the right side is congruent to 3 modulo 4.

In 1900 David Hilbert (1862–1943), in a lecture to the International Congress of Mathematicians in Paris, discussed 23 important unsolved problems in mathematics. The tenth problem on his list asks whether there exists a decision procedure for Diophantine equations. Hilbert phrased the problem, which he called "Entscheidung der Lösbarkeit einer diophantischen Gleichung" ("Determination of the Solvability of a Diophantine Equation"), as follows:

> Given a Diophantine equation with any number of unknown quantities and with rational integral numerical coefficients: *To devise a process according to which it can be determined by a finite number of operations whether the equation is solvable in rational integers.*

In this context, a Diophantine equation has integer coefficients and its solutions (if any) are integers. Hilbert's Tenth Problem calls for a universal procedure which, given any Diophantine equation, will determine whether it is solvable or unsolvable. For example, the procedure should be able to decide whether an equation such as

$$3x^5 - 7xy + 8y^3 = y^4 - 7x^2 + 100,$$

has a solution in integers x and y.

In 1973 Yuri Matiyasevich demonstrated that there is no such procedure: Hilbert's Tenth Problem is unsolvable. Matiyasevich's proof uses a Diophantine representation of the "Fibonacci function" $n = f_m$, that is, a polynomial whose integer solutions are the ordered pairs (m, n) for which $n = f_m$. In our Problem, we discussed a representation of the set of Fibonacci numbers but not of the Fibonacci function. Finding a Diophantine representation of the Fibonacci function is difficult and is the crux of Matiyasevich's proof. See [16] for a comprehensive treatment.

Perrin's Sequence

Perrin's sequence[12] $\{a_n\}$ is defined by the following recurrence formula: $a_0 = 3, a_1 = 0, a_2 = 2$, and $a_n = a_{n-3} + a_{n-2}$, for $n \geq 3$. So Perrin's sequence is

3, 0, 2, 3, 2, 5, 5, 7, 10, 12, 17, 22, 29, 39, 51, 68, 90, 119, 158, 209, ...

Find a composite number n that divides a_n. Also, prove that if p is a prime number, then p divides a_p.

Note. If it were true that n divides a_n if and only if n is prime, then we would have a polynomial-time test for primality, as we can compute the nth term of any linear recurrence modulo k in $n \log n$ time.

---

[12] The sequence is named after R. Perrin, who studied it in 1899, but the sequence was investigated in 1878 by Edouard Lucas (1842–1891).

Solution

To find a composite number n that divides a_n, we will need to compute $a_n \mod n$, as in the following algorithm.

Computing a_n mod n

Choose n.
$\{x, y, z\} \leftarrow \{3, 0, 2\}$.
Do $n - 2$ times:
  $\{x, y, z\} \leftarrow \{y, z, \mod (x + y, n)\}$.
Output z.

Which n should we try? According to the statement of the problem, it doesn't make sense to take n to be a prime. A reasonable guess is to take n to be the square of a prime, $n = p^2$.

Finding n (a prime squared) such that $n \mid a_n$

For $m = 1$ to 100 do:
  $n \leftarrow p_m^2$ (p_m is the mth prime);
  $\{x, y, z\} \leftarrow \{3, 0, 2\}$;
  Do $n - 2$ times:
    $\{x, y, z\} \leftarrow \{y, z, \mod (x + y, n)\}$;
  if $z = 0$, output n.

271441

Hence $271441 (= 521^2)$ is a number of the desired kind.

Now we prove that if p is a prime, then p divides a_p. The characteristic polynomial of the sequence is $x^3 - x - 1 = 0$, with roots, say, α, β, and γ. We find that the general term is $a_n = \alpha^n + \beta^n + \gamma^n$. The simplicity of this general term is one of the nice traits of Perrin's sequence. Let p be a prime. Then, since $\alpha + \beta + \gamma = 0$, we have $0 = (\alpha + \beta + \gamma)^p = \alpha^p + \beta^p + \gamma^p + ps$, where s is a symmetric polynomial (with integer coefficients) in α, β, γ. Hence, by the fundamental theorem of symmetric polynomials (see Bonus), s is a polynomial in the elementary symmetric polynomials in α, β, and γ. The elementary symmetric polynomials are $\alpha + \beta + \gamma$, $\alpha\beta + \beta\gamma + \gamma\alpha$, and $\alpha\beta\gamma$. These expressions are given by the coefficients of the characteristic polynomial of the recurrence relation:

$$x^3 - x - 1 = (x - \alpha)(x - \beta)(x - \gamma)$$
$$= x^3 - (\alpha\beta + \beta\gamma + \gamma\alpha)x^2 + (\alpha + \beta + \gamma)x - \alpha\beta\gamma.$$

Since the characteristic polynomial has integer coefficients, s is an integer. It follows that p divides a_p.

Bonus: The Fundamental Theorem of Symmetric Polynomials

Recall the elementary symmetric polynomials from p. 31:

$$s_1 = x_1 + x_2 + \cdots + x_n,$$

$$s_2 = \sum_{1 \leq i < j \leq n} x_i x_j,$$

$$\vdots$$

$$s_n = x_1 x_2 \ldots x_n.$$

THE FUNDAMENTAL THEOREM OF SYMMETRIC POLYNOMIALS. If f is a symmetric polynomial in the variables $x_1, x_2, \ldots, x_n$, with integer coefficients, then f is a polynomial in the elementary symmetric polynomials in $x_1, x_2, \ldots, x_n$, with integer coefficients.

For example, the symmetric function $x^3 y + x^3 z + y^3 z + y^3 x + z^3 x + z^3 y$ may be written as

$$s_1^2 s_2 - 2s_2^2 - s_1 s_3.$$

We say that the weight of a monomial $x_1^{a_1} x_2^{a_2} \ldots x_n^{a_n}$ is $(a_1, a_2, \ldots, a_n)$.

Let f be a symmetric polynomial. Order the monomials of f lexicographically according to weight. Choose a monomial of greatest weight $(a_1, a_2, \ldots, a_n)$, with $a_1 \geq a_2 \geq \cdots \geq a_n$. This monomial is "killed off" by a monomial in the elementary symmetric polynomials of weight $(a_1 - a_2, a_2 - a_3, \ldots, a_{n-1} - a_n, a_n)$, i.e., the polynomial

$$s_1^{a_1 - a_2} s_2^{a_2 - a_3} \ldots s_{n-1}^{a_{n-1} - a_n} s_n^{a_n}.$$

Subtract off this polynomial and the remainder is a symmetric polynomial of lesser weight. Continue this process until we have "killed off" the polynomial.

Let's work out the details of this procedure for the example, given earlier, of the polynomial

$$s = x^3 y + x^3 z + y^3 z + y^3 x + z^3 x + z^3 y.$$

We see that the greatest weight of a monomial is $(3, 1, 0)$. Performing the reduction indicated above, we obtain $(2, 1, 0)$. Hence, we calculate

$$s - s_1^2 s_2 = -2x^2 y^2 - 2x^2 z^2 - 2y^2 z^2 - 5xyz^2 - 5x^2 yz - 5xy^2 z.$$

We find that the greatest weight is $(2, 2, 0)$. Performing the reduction, we obtain $(0, 2, 0)$. Hence, we calculate

$$s - s_1^2 s_2 + 2s_2^2 = -x^2 yz - xy^2 z - xyz^2.$$

It's easy to see that this last expression is equal to $-s_1 s_3$. Therefore

$$s = s_1^2 s_2 - 2s_2^2 - s_1 s_3.$$

3.5 Combinatorics

Integer Triangles

How many incongruent triangles have integer side lengths and perimeter 10? There are only two: $(2, 4, 4)$ and $(3, 3, 4)$. (We specify a triangle by giving the ordered triple of its side lengths in nondecreasing order. A triple (a, b, c) must satisfy the triangle inequality $a + b > c$.)

How many triangles have integer side lengths and perimeter 10^{100}?

Solution

The number of such triangles is $2083\ldots3$, where there are one hundred ninety-six 3's. We need some theory to arrive at this number. Once we have the theory, we can calculate the number by hand!

Let $t(n)$ be the number of integer triangles of perimeter n. Let's generate some data. (It is convenient to set $t(0) = 0$.)

n	0	1	2	3	4	5	6	7	8
$t(n)$	0	0	0	1	0	1	1	2	1

The sequence $\{t(n)\}$ is known as Alcuin's sequence.[13] More about this fascinating sequence in the Bonus....

We will employ a generating function. The main idea is that we can "build up" to any triangle (a, b, c) starting from the $(1, 1, 1)$ triangle. Suppose that (a, b, c) is a triangle of perimeter n, i.e., $n = a + b + c$. How can we build up to, say, the triangle $(23, 37, 40)$? A naive approach is to add triples of the form $(1, 1, 1)$, $(0, 1, 1)$, and $(0, 0, 1)$. We can certainly obtain the desired triangle in a unique way:

$$(23, 37, 40) = (1, 1, 1) + 22(1, 1, 1) + 14(0, 1, 1) + 3(0, 0, 1).$$

The snag is that by this method we can construct triples that don't satisfy the triangle inequality, e.g., the triple $(23, 37, 100)$.

Instead, we go by steps $(0, 1, 1)$, $(1, 1, 1)$, and $(1, 1, 2)$. These triples satisfy the weak triangle inequality $a + b \geq c$, and the weak triangle inequality is good enough, since we start with a genuine triangle. A unique solution exists to the equation

$$(a, b, c) = (1, 1, 1) + \alpha(0, 1, 1) + \beta(1, 1, 1) + \gamma(1, 1, 2),$$

---

[13] Our sequence $\{t(n)\}$ is called Alcuin's sequence because it generalizes a problem given by Alcuin (735–804) in a problems book called "Propositions of Alcuin, A Teacher of Emperor Charlemagne, for Sharpening Youths." The generalized problem may be stated as follows: A dying father has n empty casks, n casks half-full of wine, and n casks full of wine that he wishes to divide among his three sons so that each receives the same number of casks and the same amount of wine. In how many ways may this be done? We met the $n = 7$ case of this problem in "Cookie Jar Division" on p. 2. In terms of cookie jars, each full jar of cookies must be counterbalanced by an empty jar, so all we need to do is specify the number of full cookie jars that each person receives (as in the diagram on p. 2). No one may have more than $n/2$ full jars, since there wouldn't be enough empty jars to compensate. So, if the three people have a, b, and c full jars, then $a + b + c = n$, where, say, $a \leq b \leq c$, and $c \leq n/2$. This means that (a, b, c) is an integer triangle that is possibly degenerate, i.e., $a + b = c$. If n is odd, the degenerate case cannot occur, so the number of solutions is $t(n)$. If n is even, then the number of solutions is $t(n + 3)$. Can you figure out why?

namely, $\alpha = b-a$, $\beta = a+b-c-1$, $\gamma = c-b$. Since $2\alpha + 3\beta + 4\gamma = a+b+c-3 = n-3$, we see that $t(n)$ is equal to the number of ways of writing $n - 3$ as a sum of 2's, 3's, and 4's (order of terms is unimportant).

The generating function that captures this counting problem is

$$\frac{x^3}{(1-x^2)(1-x^3)(1-x^4)} = t(0) + t(1)x + t(2)x^2 + t(3)x^4 + t(4)x^4 + \cdots.$$

In order to understand this, consider first the simpler generating function

$$\frac{1}{(1-x)(1-x^2)(1-x^3)}.$$

Each factor in the denominator yields a geometric series, and the product of these series is

$$(1+x+x^2+x^3+x^4+\cdots)(1+x^2+x^4+x^6+x^8+\cdots)(1+x^3+x^6+x^9+x^{12}+\cdots).$$

The first factor consists of the sum of all nonnegative powers of x, the second factor all even powers, and the third factor all powers that are multiples of three. What is the coefficient of, say, x^{10}? It is the number of ways of writing 10 as a sum of 1's (coming from the first factor), 2's (from the second factor), and 3's (from the third factor). Our generating function for the triangles of perimeter n works similarly. The allowable summands are 2, 3, and 4, accounting for the factors in the denominator. The x^3 in the numerator offsets the count by 3 so that it counts partitions of $n - 3$.

Using partial fractions (and perhaps a computer algebra system), we can rewrite the generating function as

$$-\frac{1}{24(x-1)^3} + \frac{13}{288(x-1)} - \frac{1}{16(x+1)^2} - \frac{1}{32(x+1)} - \frac{x+1}{8(x^2+1)} + \frac{x+2}{9(x^2+x+1)}.$$

Who would have thought that such a suave-looking rational function would yield such uncouth partial fractions? Although the expansion may be ugly, it will give rise to a beautiful formula.

Via the binomial theorem (for arbitrary exponents), the first four terms can be written

$$\frac{1}{24}\sum_{n=0}^{\infty}\binom{-3}{n}(-1)^n x^n - \frac{13}{288}\sum_{n=0}^{\infty} x^n - \frac{1}{16}\sum_{n=0}^{\infty}\binom{-2}{n}x^n - \frac{1}{32}\sum_{n=0}^{\infty}(-1)^n x^n.$$

Using the identity $\binom{-k}{n}(-1)^n = \binom{n+k-1}{n}$, we can simplify the coefficient of x^n to

$$\frac{1}{24}\binom{n+2}{n} - \frac{13}{288} - \frac{1}{16}(-1)^n(n+1) - \frac{1}{32}(-1)^n$$

$$= \frac{6n^2 + 18n - 1 - 18n(-1)^n - 27(-1)^n}{288}.$$

Multiplying the final two terms by $1 - x^2$ and $1 - x$, respectively, we observe that the resulting coefficients of x^n follow a pattern modulo 12. This pattern is $c/72$ where c is given in the table.

n mod 12	0	1	2	3	4	5	6	7	8	9	10	11
c	7	-17	1	25	-17	-17	25	1	-17	7	1	1

3.5 Combinatorics

Thus, we obtain the formula $t(n) = n^2/48 + (c-7)/72$, for n even, and $t(n) = (n+3)^2/48 + (c-7)/72$, for n odd. This can be represented as $t(n) = \{n^2/48\}$, for n even, and $\{(n+3)^2/48\}$, for n odd. Here, $\{x\}$ is the nearest integer to x. So, to answer our question, we compute $\{10^{200}/48\}$ by long division (observing a recurring pattern in the quotient), and obtain the result stated above.

Bonus: Alcuin's Sequence

Since Alcuin's sequence has a rational generating function, the sequence satisfies a linear recurrence relation (of order 9):

$$t(n) = t(n-2) + t(n-3) + t(n-4) - t(n-5) - t(n-6) - t(n-7) + t(n-9), \quad n \geq 9.$$

The form of the recurrence relation comes from the expanded denominator in the generating function. The initial values, for $0 \leq n \leq 8$, are given in our table of data.

Here are three neat facts about Alcuin's sequence:

1. It is a zigzag sequence (its values alternately rise and fall), for $n \geq 6$. You can prove this directly from the definition of $t(n)$ or from the formula for $t(n)$.

2. Its recurrence relation is palindromic, that is, the sequence is defined by the same recurrence relation going forward and backward. Moreover, since there are eight 0's in a row, for $-5 \leq 0 \leq 2$, the sequence is the same forwards and backwards. So if we care to define $\{t(n)\}$ for negative n, we have the curious relation $t(-n) = t(n-3)$. It follows that the explicit formula for $t(n)$ given in the Solution is valid for negative n. We aren't claiming that a triangle with negative perimeter is meaningful, only that the sequence has strange symmetry.

3. Consider the sequence modulo 2. We obtain a repeating cycle of length 24:

   0, 0, 0, 1, 0, 1, 1, 0, 1, 1, 0, 0, 1, 1, 0, 1, 1, 0, 1, 0, 0, 0, 0, 0.

   We say that the period mod 2 is 24. In general, given any modulus $m \geq 2$, the period of Alcuin's sequence mod m is $12m$. You can prove this from the formula for $t(n)$.

Topsy-Turvy Tournaments

A tournament is a complete graph in which each edge is replaced by an arrow. The direction of the edge shows who wins a game between teams designated by the two endpoint vertices.

We say that a tournament has Property k if, for every k vertices, some vertex beats all of them. For example, the tournament below has Property 1. Find a tournament with Property 2.

Solution

The following diagram shows a tournament with Property 2. It is called a quadratic residue tournament. The vertices are labeled 0 through 7, since the directions of the edges can be given by a simple rule using modulo 7 arithmetic. Vertex i is directed to vertex j if $i - j$ is a square modulo 7. The squares modulo 7 are $0^2 \equiv 0 \pmod 7$, $(\pm 1)^2 \equiv 1 \pmod 7$, $(\pm 2)^2 \equiv 4 \pmod 7$, and $(\pm 3)^2 \equiv 9 \equiv 2 \pmod 7$. Observe that a nonzero element x is a square modulo 7 if and only if $-x$ is not a non-square modulo 7, so that if $i - j$ is a square, then $j - i$ is a non-square.

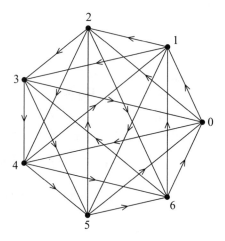

Bonus: The Probabilistic Method

It's not nearly as easy to find a tournament with, say, Property 10. Instead, we give an existence argument to show that given an arbitrary integer k there exists a tournament with Property k (without constructing it).

THEOREM (Paul Erdős). Let k be a positive integer. For some integer n, there exists a tournament on n vertices with Property k.

We employ a probabilistic approach. Let k be a given positive integer and n be an integer that we will determine later. Assume that the edges of the tournament on n vertices are directed one way or the other at random, with equal probability, and independently of all the other edges. We call such a tournament a "random tournament." For every subset S of k vertices, let A_S be the event that there is no vertex that beats all the vertices of S. In order for event A_S to occur, none of the $n - k$ vertices in the complement of S may be directed to all vertices of S. Hence

$$\Pr(A_S) = (1 - 2^{-k})^{n-k}.$$

Since probabilities are subadditive (the probability of a union is at most equal to the sum of the probabilities of the individual events), we have

$$\Pr\left(\bigcup_S A_S\right) \leq \sum_S \Pr(A_S) = \binom{n}{k}(1 - 2^{-k})^{n-k}.$$

3.5 Combinatorics

Observe that $\Pr(\bigcup_S A_S)$ is the probability that there does not exist a tournament with Property k. The expression on the right is the product of two terms: $\binom{n}{k}$ and $\left(1 - 2^{-k}\right)^{n-k}$. As k is fixed, $\binom{n}{k}$ is a polynomial in n (of degree k), while $\left(1 - 2^{-k}\right)^{n-k}$ is an exponential function in n (with base less than 1). As n tends to infinity, the exponential function dominates and the product of the two terms tends to 0. Hence, the upper bound tends to 0 and at some point must be less than 1. By taking complements, we see that for n sufficiently large, a random tournament has Property k with positive probability. But this probability is equal to the number of tournaments with Property k divided by the total number of tournaments (on n vertices). Therefore, there exists a tournament with Property k.

Following the method of the proof, we deduce that there exists a tournament with Property 10 on 102653 vertices, since

$$\binom{n}{10}(1 - 2^{-10})^{n-10} < 1,$$

for $n = 102653$.

Sudoku Solving

	1		6				8	3
				5		2	1	
	2	6		8	1			
5						7		
	8	4	2		7	5	3	
		2						8
			5	3		6	2	
	5	7		4				
9	6				2		5	

Sudoku is a Japanese word meaning "single number." The goal in a Sudoku puzzle[14] is to fill in the blank cells of a 9×9 grid so that every row, column, and 3×3 box contains every digit from 1 to 9. If you are not familiar with the puzzle, you might want to try solving the above specimen.

Can you program a computer to solve Sudoku puzzles?

---

[14] Howard Garns created the puzzle in 1979 for *Dell Pencil Puzzles & Word Games* magazine. The puzzle was called "Number Place." The puzzle was rediscovered in 1997 by Wayne Gould, who saw a partly completed puzzle, called Sudoku, in a Japanese book shop. Gould gave the puzzle to *The Times* newspaper in Britain, which first published it in 2004. The tremendous popularity of Sudoku internationally has led to it being called "the Rubik's cube of the 21st century." Besides being fun to figure out, Sudoku gives rise to some interesting mathematical questions (both solved and unsolved) [10].

Solution

There may be as many different solutions as there are computer programmers. We give a solution, by Donald Knuth, using exact covers.

What is an exact cover? Consider, for example the set

$$S = \{a, b, c, d, e, f, g\}$$

and the subsets

$$X_1 = \{c, e, f\}$$
$$X_2 = \{a, d, g\}$$
$$X_3 = \{b, c, f\}$$
$$X_4 = \{b\}$$
$$X_5 = \{b, g\}$$
$$X_6 = \{d, e, g\}.$$

Is there a collection of the X_i whose union contains each element of S exactly once? Yes, the subsets X_1, X_2, and X_4 constitute such a collection. We call such a collection an exact cover.

In general, an exact cover problem is to find a collection of sets that "cover" each element in the base set exactly once.

An exact cover problem can be given in terms of a binary matrix. Define the matrix so that its columns represent the elements of the base set and the rows represent the subsets. Put a 1 in a given position if the corresponding element is in the subset; put a 0 in the position otherwise. The binary matrix associated with our example is

$$\begin{bmatrix} 0 & 0 & 1 & 0 & 1 & 1 & 0 \\ 1 & 0 & 0 & 1 & 0 & 0 & 1 \\ 0 & 1 & 1 & 0 & 0 & 1 & 0 \\ 0 & 1 & 0 & 0 & 0 & 0 & 0 \\ 0 & 1 & 0 & 0 & 0 & 0 & 1 \\ 0 & 0 & 0 & 1 & 1 & 0 & 1 \end{bmatrix}.$$

The columns represent the elements a, b, c, d, e, f, and g, and the rows represent the subsets X_1, X_2, X_3, X_4, X_5, and X_6. An exact cover problem in the context of a binary matrix is to find a set of rows that have exactly one 1 in each column. Thus, in our example, the first, second, and fourth rows constitute a solution.

How do we convert the Sudoku problem into an exact cover problem? We form a 729 × 324 binary matrix that encodes all possible ways to put a number into the Sudoku grid. Since the grid has 81 cells and there are 9 possible numbers to put in a cell, there are $81 \cdot 9 = 729$ possible ways to put a number in the grid. This accounts for the number of rows in our matrix. Each row has exactly four 1's, corresponding to conditions that are met when a number is placed in the grid. The conditions are of four types:

- xOy means that cell (x, y) is occupied.
- xRy means that number x is in row y.

3.5 Combinatorics

- xCy means that number x is in column y.
- xBy means that number x is in block y.

These conditions are represented by the $4 \cdot 9^2 = 324$ columns.

The given numbers in a Sudoku puzzle are represented by given rows of the matrix. Solving the Sudoku puzzle is equivalent to extending the set of given rows to an exact cover of the matrix. The fact that there is exactly one 1 in each column of an exact cover is equivalent to the fact that there is exactly one i (where $1 \leq i \leq 9$) in each row, column and block of the Sudoku grid and that each cell is occupied by a number.

So we can solve a Sudoku puzzle by solving the exact cover problem that extends the given set of rows of the matrix.

Here is a depth-first search algorithm that solves the Exact Cover Problem.

Exact Cover Algorithm

```
Let M be a binary matrix.
If M is empty, then the problem is solved.
Otherwise, choose a column c with the minimum number
of 1's. If this number is zero, then terminate
unsuccessfully.
Choose, in turn, all rows r such that M[r,c] = 1.
Include r in a partial solution.
For each j such that M[r,j] = 1:
  Delete column j from M.
  For each i such that M[i,j] = 1, delete row i from M.
Repeat this algorithm recursively on the reduced
matrix M.
```

A computer implementation of the algorithm solves every Sudoku problem instantly. If there is no solution, then that is also determined instantly. For instance, the following puzzle isn't solvable because there is no way to put a 4 in the top-left block. The program discovers this right away since there is no 1 in column $4B1$.

The algorithm has a nice symmetry in that the search is over all the digits that can fit in a cell and all the places a digit can fit in a row, column, or block.

I'd like to end by relating that I once saw an advertisement for Sudoku books that said, "Don't worry—there's no math involved."

Bonus: Dancing Links

The exact cover problems that represent Sudoku puzzles are not particularly large (they have dimensions 729 × 324) and are solved instantly by a computer implementing our Exact Cover Algorithm. However, larger exact cover problems might pose a difficulty in that our algorithm calls for a lot of searching for 1's in the binary matrix. Knuth has suggested a method for efficiently implementing the algorithm (see `www-cs-faculty.stanford.edu/~knuth/preprints.html`). He calls the method Dancing Links because the data structures used are linked lists and these links execute a "dance" as the algorithm proceeds. In particular, the 1's in the matrix are nodes in circular doubly-linked lists, one such list for each column and one for each row. Each 1 is linked to its neighboring 1's to the left and right, above and below, and there is "wrap-around" at the borders of the matrix. In addition, each column is linked to a node called a column header, which keeps track of the number of 1's in that column. As the algorithm proceeds, various rows and columns are eliminated or replaced. The linked list data structure makes it much easier to keep track of all of this.

The SET Game

The SET game[15] consists of a deck of 81 cards, each displaying certain figures. The figures have four attributes: color, shape, number of objects, and shading. Each attribute has three possible values, as shown in the following table.

attribute	values
color	green, purple, red
shape	oval, diamond, squiggle
number	1, 2, 3
shading	solid, open, striped

For example, one of the cards looks like this:

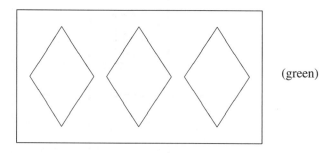

(green)

---

[15] The SET game was created in 1991 by the geneticist Marsha Falco.

3.5 Combinatorics

A "set" consists of three cards that, in each attribute, have all the same value or all different values. For example, the cards with attributes {green, oval, 3, open}, {green, diamond, 2, open}, and {green, squiggle, 1, open} are a set.

The goal in the game SET is to obtain a set. However, in this problem, we are interested in the minimum number of cards that *force* the existence of a set.

We can think of the attributes as coordinates and let each coordinate take the possible values 0, 1, and 2 (to represent the three values for that attribute). Thus, the 81 ($= 3^4$) cards are $(0, 0, 0, 0)$, $(0, 0, 0, 1)$, ..., $(2, 2, 2, 2)$. With this representation, a set consists of three ordered 4-tuples (vectors) that sum to $(0, 0, 0, 0)$, where addition is carried out modulo 3. For example, the cards $(0, 1, 1, 0)$, $(2, 2, 1, 0)$, and $(1, 0, 1, 0)$ are a set. Notice that for any two cards, there is a unique third card that makes a set with them.

The following 20 cards do not contain a set:

$(0,0,1,1)$ $(1,0,0,0)$ $(1,0,2,0)$ $(1,0,0,2)$
$(1,0,2,2)$ $(2,0,1,1)$ $(0,1,1,0)$ $(0,1,0,1)$
$(0,1,2,1)$ $(0,1,1,2)$ $(2,1,1,0)$ $(2,1,0,1)$
$(2,1,2,1)$ $(2,1,1,2)$ $(0,2,1,1)$ $(1,2,0,0)$
$(1,2,2,0)$ $(1,2,0,2)$ $(1,2,2,2)$ $(2,2,1,1)$.

We can also represent these vectors geometrically, as in the following diagram.

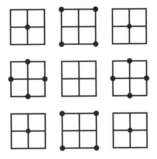

The rows of 3×3 grids represent the first coordinate and the columns of grids represent the second coordinate. In each 3×3 grid, the rows represent the third coordinate and the columns represent the fourth coordinate. The coordinates are numbered 0, 1, 2, from left to right, bottom to top. For example, the dot in the middle of the top-right grid represents the card $(2, 2, 1, 1)$. The geometric representation gives a quick portrayal of 20 cards containing no set, as we notice that no three cards going either down, across, or diagonally completely agree or disagree in each coordinate. By inspection, we can see that if we add a 21st card, then this collection will indeed contain a set. However, this is merely an example, and does not constitute a proof that *every* collection of 21 cards contains a set.

The problem is to prove that every collection of 21 cards must contain a set. Can you do this, using both mathematics and computer programming?

Solution

We can't rely on a computer alone, for the number of collections of 21 cards is $\binom{81}{21} \approx 1.4 \times 10^{19}$, too gigantic for any computer.

We need to make certain reductions first, before unleashing the power of a computer. Let's say that we have a collection of 21 cards (given as 4-tuples). Without loss of generality, we may assume that the collection contains $(0, 0, 0, 0)$. This is because, for any 21-subset, $v_1, v_2, \ldots, v_{21}$, we may take any vector, say, v_1, and subtract it from each of the vectors in the set, giving us now $(0, 0, 0, 0)$ as the first vector in the set. We are allowed to do this since, if w_1, w_2, and w_3 is a set, then so is $w_1 - v_1, w_2 - v_1$, and $w_3 - v_1$, as $(w_1 - v_1) + (w_2 - v_1) + (w_3 - v_1) = (w_1 + w_2 + w_3) - 3v_1 = (0, 0, 0, 0)$. Thus, the property of being a set is unaffected by translation. Furthermore, the set property is unaffected by multiplication by a nonsingular matrix (the proof is an exercise for the reader). Hence, we may assume that our collection contains the four standard basis vectors $(1, 0, 0, 0)$, $(0, 1, 0, 0)$, $(0, 0, 1, 0)$, and $(0, 0, 0, 1)$.

Note. We need to check that all 21 cards do not lie in a 3-dimensional subspace, but this is easy to do. A 3-dimensional subspace, say, the one consisting of all 4-tuples with fourth coordinate 0, contains 27 vectors. Assuming that the collection contains $(0, 0, 0, 0)$ and three basis vectors, we can exclude $\binom{4}{2} = 6$ other vectors, namely, those that make sets with the first four vectors. But excluding 6 vectors leaves 21 vectors, so every other vector in the subspace must be in the collection. However, this allows for numerous sets, such as $(1, 2, 1, 0), (1, 0, 1, 0), (1, 1, 1, 0)$.

Now we have five cards that are definitely in our collection, and we can exclude $\binom{5}{2} = 10$ cards that are definitely not, namely, the ones that make sets with the five included cards. At this point, the size of the problem is

$$\binom{81 - 15}{21 - 5} = \binom{66}{16} \approx 8.6 \times 10^{14}.$$

Although this number is large, we can tackle it (with a computer) by systematically putting in vectors and ruling out vectors. This verifies that every collection of 21 cards contains a set.

Bonus: A Combinatorial Problem

A neat problem generalizes our SET game problem and our earlier problem "Lots of Permutations."

The (n, d) SET game consists of n^d cards, each representing d characteristics with n choices for each characteristic. A set is a collection of n cards which, with respect to each characteristic, all agree or all disagree. The original SET game problem corresponds to $n = 3, d = 4$.

Let $f(n, d)$ be the minimum number of cards that force a set in the generalized SET game. Some values of this function, for small n and d, are $f(3, 1) = 3$, $f(3, 2) = 5$, $f(3, 3) = 10$, $f(3, 4) = 21$. In general, we have (trivially) $f(n, 1) = n$. We can also give the value of $f(n, 2)$, for any n.

PROPOSITION. For $n \geq 1$, we have $f(n, 2) = (n-1)^2 + 1$. That is, if $(n-1)^2 + 1$ cells of an $n \times n$ grid are filled, then one of the following patterns of filled cells results: n in a row, n in a column, or n with no two in the same row or column (a transversal). Furthermore, this conclusion is not guaranteed if $(n-1)^2 + 1$ is replaced by $(n-1)^2$.

3.5 Combinatorics

Let's prove this. If we fill all the cells except the last row and column, this accounts for $(n-1)^2$ cells and there is no filled row, column, or transversal.

Now assume that we fill $(n-1)^2 + 1$ cells and there is no completely filled row or column. We must show that there is a completely filled transversal. We will use an important theorem of combinatorics called Hall's Marriage Theorem.[16]

HALL'S MARRIAGE THEOREM. Let S be a collection of subsets of an n-element set X. Suppose that, for $1 \leq k \leq n$, the union of any k members of S has cardinality at least k. Then there exists, for each $s \in S$, a distinct $x \in X$ such that $x \in s$. Such a collection of x is called a *system of distinct representatives*.[17]

Note. Of course, the conditions of the theorem are necessary for the conclusion. What is not obvious is that they are sufficient.

For a proof of Hall's Marriage Theorem, see [15].

As an example, let $X = \{1, 2, 3, 4, 5\}$, and let $s_1 = \{1, 2, 3\}, s_2 = \{2, 3, 5\}, s_3 = \{4, 5\}$, and $s_4 = \{3, 5\}$. Then one system of distinct representatives (there is more than one in this case) is $1 \in s_1, 2 \in s_2, 5 \in s_3$, and $3 \in s_4$.

To continue with our proof, let X be the columns of the grid and S the collection of filled cells in each row. We will show that the hypothesis of Hall's Marriage Theorem is satisfied, and it will therefore follow that the grid contains a filled transversal.

If the hypothesis fails for $k = 1$, then some row is empty. But then the $(n-1)^2 + 1$ filled cells occur in $n-1$ rows, and it follows that one of these rows is complete, contradicting our assumption. Hence, the hypothesis is satisfied for $k = 1$. A similar argument shows that the hypothesis is satisfied for $k = n$.

Suppose that the hypothesis fails for some k such that $1 < k < n$. Then a union of some k rows contains filled cells in at most $k-1$ columns, accounting for at most $k(k-1)$ filled cells. But then the remaining $n-k$ rows, which contain at most $(n-k)(n-1)$ filled cells, must account for the rest. However, we can show that

$$(n-k)(n-1) + k(k-1) \leq (n-1)^2,$$

for this inequality is equivalent to our assumption that $n \geq k + 1$. The contradiction means that the hypothesis of Hall's Marriage Theorem is satisfied, and we have our desired conclusion.

In general, the calculation of $f(n, d)$ is an unsolved problem. In particular, no formula is known for $f(n, 3)$.

Girth Five Graphs

The *girth* of a graph is the length of a smallest cycle in the graph. For example, the girth of the graph on the left is 5, as the graph is a 5-cycle. The girth of the graph on the right is 3, as the graph contains a triangle.

---

[16] Hall's Marriage Theorem is credited to algebraist Philip Hall (1904–1982).
[17] The whimsical name of the theorem derives from thinking of the elements of X as men and the elements of S as lists of men whom various women would like to marry. The conclusion of the theorem is a satisfactory pairing of the men with the women.

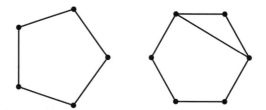

A graph is r-regular if every vertex is adjacent to exactly r vertices. The graph on the left is 2-regular and the graph on the right is not regular.

Suppose that a graph is r-regular, for some r, and has girth 5. How many vertices will the graph have? Well, every vertex is adjacent to r vertices, and none of these vertices can be adjacent to each other (or else the graph would contain a triangle). So we have the following picture.

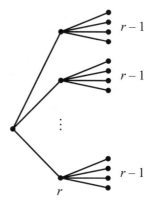

The vertices in the third column must be distinct (or else the graph would contain a 4-cycle). Hence, the graph contains at least $1 + r + r(r-1) = r^2 + 1$ vertices.

The question is, what are the possible values of r such that an r-regular graph of girth 5 contains exactly $r^2 + 1$ vertices? For example, the 5-cycle pictured above satisfies the condition where $r = 2$.

Solution

Surprisingly, only four values of r are possible: 2, 3, 7, and 57. How do we prove such a thing? The aha! idea is to introduce an adjacency matrix. Define $A = [a_{ij}]$, an $n \times n$ matrix with $n = r^2 + 1$, where $a_{ij} = 1$ if vertices v_i and v_j are adjacent and 0 otherwise. Note that A is a symmetric matrix.

The critical observation is that

$$A^2 + A - (r-1)I = J,$$

where I is the $n \times n$ identity matrix and J is the $n \times n$ matrix of all 1's. To prove this, we observe that

$$(A^2)_{ij} = \sum_{k=1}^{n} A_{ik} A_{kj}$$

$$= \text{the number of neighbors common to the vertices } v_i \text{ and } v_j.$$

3.5 Combinatorics

Now, consider two vertices, v_i and v_j of the graph. If $i = j$, then $(A^2)_{ij} = r$, and the left side of the identity to be proved counts $r + 0 - (r - 1) = 1$, which is the same as the right side counts. If $i \neq j$ and v_i and v_j are adjacent, then v_i and v_j have no common neighbors (or else the graph would have a triangle), so the left side counts $0 + 1 - (r - 1)0 = 1$, which agrees with the right side. If $i \neq j$ and v_i and v_j are nonadjacent, then v_i and v_j have exactly one common neighbor (why?), and so the left side counts $1 + 0 - (r - 1)0 = 1$, which agrees with the right side. Hence, the relation is true.

Next, we investigate the eigenvalues of A. As these numbers must be integers, there will be serious restrictions on the possible values of r. Because A is a symmetric matrix, it has n eigenvectors. In fact, A has an orthogonal basis of eigenvectors (this is known as the "spectral theorem"). One of the eigenvectors of A is the all 1's vector, with eigenvalue r. Suppose that λ is an eigenvalue of another eigenvector of A, say, e. Since the basis is orthogonal, we have $Je = 0$. Therefore

$$0 = Je$$
$$= (A^2 + A - (r-1)I)e$$
$$= \lambda^2 e + \lambda e - (r-1)e.$$

This implies that

$$\lambda^2 + \lambda - (r-1) = 0.$$

Using the quadratic formula, we find that

$$\lambda = \frac{1}{2}(-1 \pm \sqrt{4r-3}).$$

Define $\lambda_1 = \frac{1}{2}(-1 - \sqrt{4r-3})$ and $\lambda_2 = \frac{1}{2}(-1 + \sqrt{4r-3})$. Now we know that all possible eigenvalues of A are r, λ_1, and λ_2.

What can we say about the multiplicities of these eigenvalues? The eigenvalue r has multiplicity 1, since neither λ_1 or λ_2 can equal r (check). Let m_1 and m_2 be the multiplicity of λ_1 and λ_2, respectively. Hence

$$1 + m_1 + m_2 = n = r^2 + 1.$$

We also know from linear algebra that the sum of the eigenvalues of a matrix is equal to the trace of the matrix. All the diagonal elements of A are 0, so the trace of A is 0, and hence

$$r + m_1 \lambda_1 + m_2 \lambda_2 = 0.$$

Combining these last two equations together with the values of λ_1 and λ_2, we obtain

$$2r - r^2 + (m_1 - m_2)\sqrt{4r-3} = 0.$$

Consider two cases. If $m_1 - m_2 = 0$, then $r = 2$. If $m_1 - m_2 \neq 0$, then $\sqrt{4r-3}$ is a rational number and hence an integer. Denote this integer by s. Then $r = (s^2 + 3)/4$. Simplifying our equation yields

$$[s^3 - 2s - 16(m_1 - m_2)]s = 15.$$

It follows that s divides 15, and hence s is 1, 3, 5, or 15. But $s = 1$ implies that $r = 1$, which doesn't give a girth five graph. Therefore, r is 2 (from the first case), 3, 7, or 57. This result was proved in 1960 by Alan Hoffman and Robert Singleton.

Do r-regular, girth five graphs with $r^2 + 1$ vertices exist for these values of r? Our 5-cycle above is an example with $r = 2$. A little experimentation yields a graph with $r = 3$. It has ten vertices and is called the *Petersen graph*. It was discovered by Julius Petersen (1839–1920).

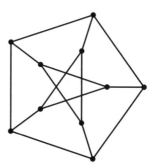

The existence of such a graph with $r = 7$ is shown in the Bonus. No one knows whether a 57-regular, girth five graph with 3250 vertices exists. Such a graph would have 3250 vertices.

Bonus: The Hoffman–Singleton Graph

We will show how to make the Hoffman–Singleton graph. Remember that the graph is to be a 7-regular graph with 50 vertices and girth 5. Let P_0, P_1, P_2, P_3, and P_4 be 5-cycles, with vertices numbered 0, 2, 4, 1, and 3 (in cyclic order). Let Q_0, Q_1, Q_2, Q_3, and Q_4 be 5-cycles, with vertices numbered 0, 1, 2, 3, and 4 (in cyclic order). The P cycles and Q cycles have altogether 50 vertices, and these are the vertices of our graph. Each vertex in one of the P cycles will be joined to exactly one vertex in each of the Q cycles. Thus, each vertex will be adjacent to $2 + 5 = 7$ vertices. Here is the recipe for how to join the vertices. A vertex i in P_j is joined to vertex $i + jk$ in Q_k, where the calculation is performed modulo 5.

The automorphism group (the group of symmetries) of the Hoffman–Singleton graph has order $252000 = 50 \cdot 7!$. Each vertex of the graph may be moved to any vertex (that is to say, the symmetry group is transitive). Once this operation is carried out, the seven vertices adjacent to any given vertex may be permuted arbitrarily. Altogether, this accounts for $50 \cdot 7!$ symmetries.

What is the symmetry group of the Petersen graph?

Unlabeled Graphs

How many graphs are there on the vertex set $\{1, 2, 3, \ldots, n\}$? There are $2^{\binom{n}{2}}$, as there are $\binom{n}{2}$ possible edges and each edge can be included or not included in the graph. For example, there are $2^{\binom{4}{2}} = 2^6 = 64$ labeled graphs on 4 vertices. However, if we regard the vertices

3.5 Combinatorics

as indistinguishable (in other words, the graph is unlabeled), then there are only 11 graphs on 4 vertices.

Here are the 11 unlabeled graphs of order 4.

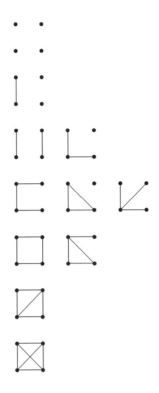

Let $\overline{g}_n$ be the number of labeled graphs on n vertices and g_n the number of unlabeled graphs on n vertices. As we have said, $\overline{g}_n = 2^{\binom{n}{2}}$. Determine a formula for g_n.

Solution

The 64 labeled graphs on 4 vertices are partitioned into 11 sets of equivalent unlabeled graphs (the graphs pictured above). For example, the 12 labeled graphs below are all equivalent as unlabeled graphs.

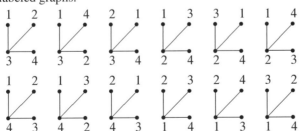

These 12 labeled graphs constitute an *orbit*. To test your understanding, determine the size of the orbit corresponding to each of the 11 unlabeled graphs above. The sum of these orbit sizes will be 64.

We need a formula for the number of orbits when a group (the group of permutations) acts on a set (the set of unordered pairs of vertices). The formula is given by Burnside's lemma. For a proof of Burnside's lemma, which is a nice example of double-counting, see [6].

BURNSIDE'S LEMMA. Let a finite group G act on a finite set X. Then the number of orbits is given by

$$\frac{1}{|G|} \sum_{g \in G} f_g,$$

where f_g is the number of elements of X fixed by g.

Let's use Burnside's lemma to prove that there are 11 unlabeled graphs of order 4. Let $V = \{1, 2, 3, 4\}$, the set of vertices of the graph, and $V^{(2)} = \{12, 13, 14, 23, 24, 34\}$, the set of unordered pairs of vertices. What is the permutation of $V^{(2)}$ given by, for example, the permutation $(1, 2)(3, 4)$ of V? Looking at what happens to each element of the unordered pairs, we find that the corresponding permutation of $V^{(2)}$ is $(12)(13, 24)(14, 23)(34)$. The number of fixed points of this permutation is 2 raised to the number of cycles, i.e., 2^4, as each cycle in a fixed point is a cycle of edges or non-edges.

The cycle lengths of a permutation make up the *cycle type* of the permutation. For example, the cycle type of the permutation $(abc)(de)(f)(ghi)$ is given as $1 + 2 + 3 + 3$. We also represent a cycle type as a vector $(c_1, c_2, \ldots, c_n)$, where c_i is the number of cycles of length i, for $1 \leq i \leq n$. The permutation $(abc)(de)(f)(ghi)$ is represented as $(1, 1, 2, 0, 0, 0, 0, 0, 0)$. Permutations of the same cycle type yield the same number of fixed points. The following table lists the cycle type of permutations of V, the number of permutations of each cycle types, the cycle types of the corresponding permutations of $V^{(2)}$, and the number of fixed points.

cycle type of permutation of V	# of permutations of given cycle type	cycle type of permutation of $V^{(2)}$	# of fixed points
$1+1+1+1$	1	$1+1+1+1+1+1$	2^6
$1+1+2$	6	$1+1+2+2$	2^4
$2+2$	3	$1+1+2+2$	2^4
$1+3$	8	$3+3$	2^2
4	6	$2+4$	2^2

Thus, by Burnside's lemma, the number of nonisomorphic graphs of order 4 is

$$\frac{1}{24}(1 \cdot 2^6 + 6 \cdot 2^4 + 3 \cdot 2^4 + 8 \cdot 2^2 + 6 \cdot 2^2) = 11.$$

The general formula, due to George Pólya,[18] is given by

$$g_n = \frac{1}{n!} \sum_c h(c) 2^{q(c)},$$

where $c = (c_1, \ldots, c_n)$ is the cycle type of a permutation of V,

$$h(c) = \frac{n!}{\prod_{k=1}^n k^{c_k} c_k!}$$

---
[18] George Pólya (1887–1985) was a master problem solver, well known for his book *How to Solve It* [17].

3.5 Combinatorics

is the number of permutations of cycle type c, and

$$q(c) = \sum_k \left\lfloor \frac{k}{2} \right\rfloor c_k + \sum_k k \binom{c_k}{2} + \sum_{r<s} \gcd(r,s) c_s c_s$$

is the number of cycles in a corresponding permutation of $V^{(2)}$. The first term of $q(c)$ comes from two vertices in the same cycle of the permutation of V. The second term comes from two vertices in different cycles of the same length. The third term comes from two vertices in different cycles of different lengths.

Bonus: Asymmetric Graphs

The typical unlabeled graph has no nontrivial symmetries and hence is an *asymmetric graph*. That is,

$$g_n \sim \frac{\bar{g}_n}{n!}.$$

I leave it to you to explain the trivial lower bound,

$$\frac{1}{n!} 2^{\binom{n}{2}} \leq g_n.$$

The complications in the formula for g_n melt away in the following proof of the upper bound. Assume that the permutation has $n - j$ fixed points, where $0 \leq j \leq n$. The case $j = 0$ gives the term $2^{\binom{n}{2}}/n!$. We will show that the other terms are bounded above by expressions which, upon division by $2^{\binom{n}{2}}/n!$, tend to 0 as $n \to \infty$.

We have

$$q(c) \leq \sum_k \left[\frac{k}{2} c_k + \frac{k}{2} c_k (c_k - 1) \right] + \sum_{r<s} \min(r,s) c_r c_s$$

$$\leq \frac{1}{2} \sum_k k c_k^2 + \sum_{r<s} \left(\frac{r+s}{2} \right) c_r c_s$$

$$= \frac{1}{2} \sum_k k c_k \sum_k c_k$$

$$= \frac{1}{2} n \sum_k c_k.$$

The number of permutations with $n - j$ fixed points is

$$\leq \binom{n}{n-j} j! = \frac{n!}{(n-j)!} \leq n^j = 2^{j \log_2 n}.$$

The case $j = 2$ yields the exponent

$$q(c) = 1 + \binom{n-2}{2} + 1 \cdot (n-2) \cdot 1 = \binom{n}{2} - n + 2.$$

Upon dividing by $2^{\binom{n}{2}}/n!$, the contribution is bounded by $2^{-n+2+2\log_2 n}$, which tends to 0 as $n \to \infty$.

Now let's look at the case $j \geq 3$, so that $1 - j/2 < 0$. We obtain
$$\sum_k c_k \leq n - j + j/2 = n - j/2$$
and
$$q(c) \leq \frac{1}{2}n(n - j/2) = \binom{n}{2} + \frac{1}{2}n(1 - j/2).$$

Upon dividing by $2^{\binom{n}{2}}/n!$, the contribution is bounded by $2^{\frac{1}{2}n(1-j/2)+j \log_2 n}$, which tends to 0 as $n \to \infty$.

We conclude that
$$g_n \sim \frac{2^{\binom{n}{2}}}{n!}.$$

A
Toolkit

AM, GM, HM and AM–GM–HM inequalities. Let $x_1, \ldots, x_n$ be positive real numbers. The *arithmetic mean* (AM), *geometric mean* (GM), and *harmonic mean* (HM) of x_1, $\ldots, x_n$ are defined as

$$\text{AM} = \frac{1}{n} \sum_{i=1}^{n} x_i,$$

$$\text{GM} = \left(\prod_{i=1}^{n} x_i \right)^{1/n},$$

$$\text{HM} = \frac{n}{\sum_{i=1}^{n} \frac{1}{x_i}}.$$

These means satisfy the inequalities

$$\text{HM} \leq \text{GM} \leq \text{AM},$$

with equality if and only if all the x_i are equal.

Area. The area of a square of side length s is s^2. The area of a triangle of base b and height h is $bh/2$ (also see Heron's formula under the entry for Triangle). The area of a parallelogram of base b and height h is bh. The area of a circle of radius r is πr^2.

Binomial coefficient. The expression

$$\binom{n}{k} = \frac{n!}{k!(n-k)!},$$

for $0 \leq k \leq n$, which is equal to the number of k-element subsets of an n-element set.

183

Another important binomial coefficient expression is
$$\binom{n+k-1}{k-1},$$
which is equal to the number of distributions of n indistinguishable objects into k classes.

Binomial theorem. The identity
$$(x+y)^n = \sum_{k=0}^{n} \binom{n}{k} x^k y^{n-k}, \quad n \geq 0.$$
Letting $x = y = 1$, we obtain the identity
$$2^n = \sum_{k=0}^{n} \binom{n}{k}.$$

Chinese remainder theorem. If $n_1, n_2, \ldots, n_k$ are pairwise relatively prime positive integers and $r_1, r_2, \ldots, r_k$ are any integers, then there exists an integer x satisfying the simultaneous congruences
$$x \equiv r_1 \pmod{n_1},$$
$$x \equiv r_2 \pmod{n_2},$$
$$\vdots$$
$$x \equiv r_k \pmod{n_k}.$$
Furthermore, x is unique modulo $n_1 n_2 \ldots n_k$.

We can find the value of x as follows. Let $n = n_1 n_2 \ldots n_k$. For $1 \leq i \leq k$, solve the congruence $(n/n_i) x_i \equiv r_i \pmod{n_i}$. Then
$$x \equiv \sum_{i=1}^{k} \frac{n}{n_i} x_i \pmod{n}.$$

Combination. A selection of objects from a set in which the order of the objects is unimportant. The number of combinations of size k from a set of size n is given by the binomial coefficient $\binom{n}{k}$.

Conic sections. The intersection of a plane with a double cone. The nondegenerate conic sections are circles, parabolas, ellipses, and hyperbolas. After suitable translations and rotations, the equation of a conic section can be put into a simple form. The equation of a circle is
$$x^2 + y^2 = r^2.$$
The equation of a parabola is
$$y = kx^2.$$

The equation of an ellipse is
$$\frac{x^2}{a^2} + \frac{y^2}{b^2} = 1.$$
The equation of a hyperbola is
$$\frac{x^2}{a^2} - \frac{y^2}{b^2} = 1.$$

Convex function. A real-valued function f is convex if
$$f((1-\lambda)x + \lambda y) \leq (1-\lambda)f(x) + \lambda f(y),$$
for all real x, y and $0 \leq \lambda \leq 1$.

Derivative. The derivative of a real-valued function gives the rate of change of the function. An important rule of differentiation is the power rule: the derivative of the function $f(x) = x^n$ is $f'(x) = nx^{n-1}$.

Descartes' rule of signs. The number of positive roots of a polynomial is equal to the number of sign changes of its coefficients, or less by an even number.

Determinant. The determinant of a square matrix $A = [a(i,j)]_{n \times n}$ is given by
$$\det A = \sum_\sigma (-1)^{\text{sgn}(\sigma)} \prod_{i=1}^n a(i, \sigma(i)),$$
where the sum is over all permutations σ of $\{1, 2, \ldots, n\}$ and $\text{sgn}(\sigma)$, the sign of σ, is $+1$ if σ is an even permutation and -1 if σ is an odd permutation. The determinant of a matrix is nonzero if and only if the matrix is invertible.

Directed graph. A graph in which each edge $\{x, y\}$ has a direction, i.e., from x to y or from y to x.

Euler's formula. Suppose that a connected planar graph with V vertices and E edges partitions the plane into F regions ("faces"). Then
$$V + F = E + 2.$$

Euler's function. For $n \geq 1$, let $\phi(n)$ be the number of integers between 1 and n that are relatively prime to n. If the prime factorization of n is
$$n = \prod_{i=1}^k p_i^{e_i},$$
then
$$\phi(n) = \prod_{i=1}^k \left(p_i^{e_i} - p_i^{e_i - 1} \right).$$
The following fact is useful:
$$\sum_{d \mid n} \phi(d) = n.$$

Fermat's (little) theorem. If p is a prime number and $p \nmid a$, then

$$a^{p-1} \equiv 1 \pmod{p}.$$

Fibonacci sequence. The sequence

$$\{f_n\} = \{0, 1, 1, 2, 3, 5, 8, 13, 21, 34, \ldots\}$$

defined by $f_0 = 0$, $f_1 = 1$, and $f_n = f_{n-1} + f_{n-2}$, for $n \geq 2$.

In order to find an explicit formula for the Fibonacci numbers, assume that there exists a value of x such that the sequence $\{x^n\}$ satisfies the recurrence relation. Then

$$x^n = x^{n-1} + x^{n-2},$$

and hence

$$x^2 = x + 1,$$

or

$$x^2 - x - 1 = 0.$$

The polynomial $x^2 - x - 1$ is called the *characteristic polynomial* of the recurrence relation. From the quadratic formula, we find that the roots of the characteristic polynomial are

$$\phi = \frac{1 + \sqrt{5}}{2}, \quad \hat{\phi} = \frac{1 - \sqrt{5}}{2}.$$

Hence, the sequences

$$\{\phi^n\}, \quad \{\hat{\phi}^n\}$$

satisfy the recurrence relation. If two sequences u and v satisfy the recurrence relation, then so does any linear combination of them, $c_1 u + c_2 v$, with c_1 and c_2 constants. Hence, the sequence

$$\{c_1 \phi^n + c_2 \hat{\phi}^n\}$$

satisfies the Fibonacci recurrence relation. From the initial values, we deduce that

$$c_1 = \frac{1}{\sqrt{5}}, \quad c_2 = -\frac{1}{\sqrt{5}},$$

and so the sequence

$$\left\{ \frac{\phi^n - \hat{\phi}^n}{\sqrt{5}} \right\}$$

satisfies the recurrence relation and the initial conditions. But there is only one sequence that does both. Therefore

$$f_n = \frac{\phi^n - \hat{\phi}^n}{\sqrt{5}}, \quad n \geq 0.$$

Field. A *field* F is a set (with at least two elements) on which are defined two binary operations, $+$ and $\cdot$, such that the following conditions hold:

1. F is an abelian group with respect to $+$;

2. $F - \{0\}$ (where 0 is the additive identity) is an abelian group with respect to $\cdot$;
3. (distributive law) for all $x, y, z \in F$,
$$x \cdot (y + z) = x \cdot y + x \cdot z.$$

Examples: (1) the set $\mathbf{R}$ of real numbers with the usual addition and multiplication; (2) the set $\mathbf{Q}$ of rational numbers with the usual addition and subtraction; (3) the set $\mathbf{Z}_2 = \{0, 1\}$ with addition and multiplication modulo 2.

A *finite field* exists of any order p^k, where p is a prime and $k \geq 1$. These are the only possible orders of finite fields.

Fundamental theorem of arithmetic. Every integer $n > 1$ has a unique (up to order of factors) prime factorization
$$n = p_1^{e_1} p_2^{e_2} \cdots p_k^{e_k},$$
where the p_i are distinct primes and the e_i are positive integers.

Generating function. The (ordinary) generating function of a sequence $\{a_n\}$ is the power series
$$a_0 + a_1 x + a_2 x^2 + a_3 x^3 + a_4 x^4 + \cdots.$$
For example, the generating function for the Fibonacci sequence is
$$x + x^2 + 2x^3 + 3x^4 + 5x^5 + 8x^6 + 13x^7 + \cdots.$$

Golden ratio. The number $\phi = (1 + \sqrt{5})/2$. This number has the property that
$$\phi^2 = \phi + 1.$$

Graph. A *graph* G consists of a set V of *vertices* and a set E of *edges* joining some pairs of vertices. Vertices joined by an edge are *adjacent*. Vertices not joined by an edge are *nonadjacent*. In a drawing of a graph, adjacent vertices may be joined by a straight or curved line and the lines may cross arbitrarily.

The picture below shows a graph with seven vertices and ten edges.

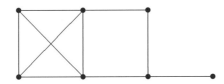

The *degree* of a vertex is the number of vertices adjacent to it. In the above graph, the vertices have degrees 1, 2, 3, 3, 3, 4, 4.

The *complete graph* on n vertices, denoted K_n, is a set of n vertices and all possible edges between vertices. The number of edges is $\binom{n}{2}$.

The *complete bipartite graph* on m, n vertices, denoted $K_{m,n}$, is a set of m vertices, a set of n vertices, and all possible edges between vertices in different sets. The number of edges is mn.

Greatest common divisor. Let a and b be integers, not both 0. The *greatest common divisor* of a and b, written $\gcd(a, b)$, is the greatest integer that divides both a and b. Given integers a and b with greatest common divisor g, there exist integers x and y such that
$$g = ax + by.$$

Group. A *group* G is a nonempty set on which is defined a binary operation $*$ satisfying the following three axioms:

1. (associativity) for all $x, y, z \in G$, $x * (y * z) = (x * y) * z$;
2. (identity element) G contains an element e with the property that, for all $x \in G$, $x * e = e * x = x$;
3. (inverse elements) for every $x \in G$ there exists an $x^{-1} \in G$ with the property that $x * x^{-1} = x^{-1} * x = e$.

Examples: (1) the set of integers $\mathbf{Z}$ under addition; (2) the set of nonzero real numbers $\mathbf{R} \setminus \{0\}$ under multiplication; (3) the set of invertible $n \times n$ matrices with real entries under matrix multiplication.

We say that G is *abelian* if $x * y = y * x$ for all $x, y \in G$; otherwise, G is *nonabelian*. Examples: the groups (1) and (2) above are abelian and (3) is nonabelian.

Group action. Let G be a group and X a nonempty set. Then a group action of G on X is a function from $G \times X$ to X such that (1) $ex = x$, for all $x \in X$, where e is the identity element of G, and (2) $g(hx) = (gh)x$, for all $g, h \in G$ and $x \in X$. In a group action, each element of G permutes the elements of X.

Integral. An indefinite integral is the antiderivative of a function. A definite integral can represent the area under a curve. An important integration rule is the power rule:
$$\int x^n \, dx = \frac{1}{n+1} x^{n+1} + C, \quad n \neq -1.$$

Intermediate value theorem. If f is a continuous real-valued function defined on the closed interval $[a, b]$, and y is any number between $f(a)$ and $f(b)$, then there exists a number x in (a, b) such that $f(x) = y$.

Jordan curve theorem. If C is a simple closed curve in the plane, then C divides the plane into two components.

Lagrange's theorem. The order of a subgroup of a finite group divides the order of the group.

Lattice point. An ordered n-tuple $(x_1, \ldots, x_n) \in \mathbf{R}^n$, where $x_1, \ldots, x_n$ are integers.

Law of cosines. Given $\triangle ABC$ labeled as below,
$$c^2 = a^2 + b^2 - 2ab \cos C.$$

A Toolkit

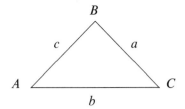

Law of sines. Given $\triangle ABC$ labeled as above,
$$\frac{\sin A}{a} = \frac{\sin B}{b} = \frac{\sin C}{c}.$$

Least common multiple. Let a and b be integers, not both 0. The *least common multiple* of a and b, written $\mathrm{lcm}(a, b)$, is the least positive integer that is a multiple of a and b.

L'Hôpital's rule. Suppose that f and g are differentiable functions and g' is nonzero on an interval containing a (except possibly at a). If $\lim_{x \to a} f(x) = 0$ and $\lim_{x \to a} g(x) = 0$, or if $\lim_{x \to a} f(x) = \pm\infty$ and $\lim_{x \to a} g(x) = \pm\infty$, then
$$\lim_{x \to a} \frac{f(x)}{g(x)} = \lim_{x \to a} \frac{f'(x)}{g'(x)}$$
(if the second limit exists).

Modulo n. Two integers a and b are congruent modulo n, and we write
$$a \equiv b \pmod{n},$$
when n divides $a - b$. For example, $7 \equiv 107 \pmod{10}$.

Multinomial coefficient. The expression
$$\binom{n}{\alpha_1, \alpha_2, \ldots, \alpha_k} = \frac{n!}{\alpha_1! \alpha_2! \ldots \alpha_k!},$$
where all $\alpha_i \geq 0$ and $\alpha_1 + \alpha_2 + \cdots + \alpha_k = n$, which is equal to the number of permutations of n elements in which there are α_i elements of type i, for $1 \leq i \leq k$.

Multinomial theorem. The identity
$$(x_1 + x_2 + \cdots + x_k)^n = \sum_{\alpha_1 + \alpha_2 + \cdots + \alpha_k = n} \binom{n}{\alpha_1, \alpha_2, \ldots, \alpha_k} x_1^{\alpha_1} x_2^{\alpha_2} \ldots x_k^{\alpha_k}, \quad n \geq 0.$$

Numbers. The set of natural numbers $\{1, 2, 3, \ldots\}$ is denoted by **N**. The set of integers $\{\ldots, -3, -2, -1, 0, 1, 2, 3, \ldots\}$ is denoted by **Z**.

The set of real numbers is denoted by **R**. A real number is *rational* if it is of the form p/q, where p and q are integers and $q \neq 0$. The set of rational numbers is denoted by **Q**. A real number that is not rational is called *irrational*. Examples: $2/3$, $-4/7$, and 14 are rational numbers, while $\sqrt{2}$, e, and π are irrational numbers.

The set of complex numbers is denoted by **C**.

A complex number is *algebraic* if it is the root of a polynomial with integer coefficients. A number that is not algebraic is called *transcendental*. Examples: $\sqrt{2}$ is algebraic and π and e are transcendental.

Pascal's identity. The identity

$$\binom{n+1}{k} = \binom{n}{k} + \binom{n}{k-1}, \quad 1 \le k \le n.$$

Pascal's triangle. Pascal's identity is the basis for generating Pascal's triangle of binomial coefficients.

```
              1
            1   1
          1   2   1
        1   3   3   1
      1   4   6   4   1
    1   5  10  10   5   1
              ⋮
```

Permutation. A selection of objects from a set in which the order of the objects is important. The number of permutations of size k from a set of size n is $n(n-1)(n-2)\cdots(n-k+1)$. If $k = n$, this number is given by the "factorial" $n! = n(n-1)(n-2)\cdots 3\cdot 2\cdot 1$.

The cycle form of a permutation shows the image of each element and the resulting cycles. For example, the permutation of $\{1, 2, 3, 4, 5, 6\}$ that sends 1 to 4, 2 to 5, 3 to 1, 4 to 3, 5 to 2, and 6 to 6 is written in cycle form as $(1, 4, 3)(5, 2)(6)$.

Pigeonhole principle. There are many versions of this useful combinatorial principle. Here is a simple version: If $n + 1$ objects are in n classes, then one of the classes must contain at least two of the objects.

Polynomial. A polynomial of degree n is an expression of the form

$$a_n x^n + a_{n-1} x^{n-1} + \cdots + a_1 x + a_0.$$

A root of a polynomial $P(x)$ is a number r such that $P(r) = 0$.

A simple relationship exists between the roots of a polynomial and its factors: r is a root if and only if $x - r$ is a factor.

Every polynomial of degree n has n roots in the field of complex numbers **C**.

Power set. The power set of a set S, denoted $\mathcal{P}(S)$, is the collection of all subsets of S. If S has n elements, then $\mathcal{P}(S)$ has 2^n elements.

Prime and composite numbers. A natural number $n \ge 2$ is *composite* if n has positive factors other than 1 and n. Otherwise, n is *prime*. Examples: 15, 100, and 2^{100} are composite numbers; 2, 17, and 1999 are primes.

A Toolkit

Principle of inclusion and exclusion. Suppose that $A_1, \ldots, A_n$ are finite sets. For $1 \leq k \leq n$, define
$$I_{n,k} = \sum_{1 \leq i_1 < \cdots < i_k \leq n} |A_{i_1} \cap \cdots \cap A_{i_k}|.$$
Then
$$|A_1 \cup \cdots \cup A_n| = \sum_{k=1}^{n} (-1)^{k+1} I_{n,k}.$$
The case $n = 2$ is the well-known Venn diagram rule $|A \cup B| = |A| + |B| - |A \cap B|$.

Ptolemy's theorem. A quadrilateral of side lengths a, b, c, and d (in cyclic order) and diagonal lengths m and n is cyclic (the vertices of the quadrilateral lie on a circle) if and only if
$$ac + bd = mn.$$

Pythagorean theorem. The side lengths of a triangle, a, b, c, satisfy the relation
$$a^2 + b^2 = c^2$$
if and only if the triangle is a right triangle with right angle opposite the side of length c.

Quadratic formula. The quadratic equation $ax^2 + bx + c = 0$ has solutions
$$x = \frac{-b \pm \sqrt{b^2 - 4ac}}{2a}.$$

Rational root theorem. If $P(x) = a_n x^n + a_{n-1} x^{n-1} + \cdots + a_1 x + a_0$ is a polynomial with integer coefficients, and p/q is a rational root of P (with $\gcd(p, q) = 1$), then p divides a_0 and q divides a_n.

Series. A series is a sum of a sequence of numbers. The number of terms in the sum may be finite or infinite.

An arithmetic series is one whose terms increase by a constant amount. Thus, an infinite arithmetic series is of the form
$$a + (a + d) + (a + 2d) + (a + 3d) + \cdots,$$
where a is the first term and d is the common difference.

A geometric series is one whose terms change by a fixed ratio. Thus, an infinite geometric series is of the form
$$a + ar + ar^2 + ar^3 + \cdots,$$
where a is the first term and r is the common ratio.

Stirling's approximation:
$$n! \sim n^n e^{-n} \sqrt{2\pi n}.$$

Triangle. The area of a triangle is $A = \frac{1}{2}ah$, where a is the length of one of the sides and h is the length of the altitude from the opposite vertex. The area is also given by *Heron's formula*:
$$A = \sqrt{s(s-a)(s-b)(s-c)},$$
where a, b, c are the side lengths and $s = \frac{1}{2}(a+b+c)$ is the *semiperimeter*.

In any triangle with side lengths a, b, and c, the *triangle inequality* asserts that
$$a < b + c.$$

If the sides are given by vectors x, y, and $x + y$, the inequality becomes
$$|x + y| \leq |x| + |y|.$$

Equality occurs if and only if the vectors are parallel (and the triangle degenerate).

A triangle's *circumcenter* is the center of its circumscribed circle. It is the intersection of the perpendicular bisectors of the sides of the triangle. The triangle's *incenter* is the center of its inscribed circle. It is the intersection of the three angle bisectors of the triangle. The *centroid* is the center of mass of the triangle. It is the intersection of the three medians of the triangle (a *median* is a line joining a vertex to the midpoint of the opposite side). The *orthocenter* is the intersection of the three altitudes of the triangle (an *altitude* is a line passing through a vertex and perpendicular to the opposite side).

Tournament. A complete directed graph, i.e., one in which every pair of vertices x, y are joined by an arrow from x to y or from y to x.

Vector space. A *vector space* V over a field F is an additive abelian group together with a rule that assigns to every $f \in F$ and $v \in V$ an element $f \cdot v \in V$ such that, for all $f, f_1, f_2 \in F$ and $v, v_1, v_2 \in V$, the following conditions hold:

1. $f \cdot (v_1 + v_2) = f \cdot v_1 + f \cdot v_2$;
2. $(f_1 + f_2) \cdot v = f_1 \cdot v + f_2 \cdot v$;
3. $f_1 \cdot (f_2 \cdot v) = (f_1 f_2) \cdot v$;
4. $1 \cdot v = v$, where 1 is the multiplicative identity of F.

The elements of V are called *vectors* and the elements of F are called *scalars*.

Examples: the vector space $\mathbf{R}^2$ over the field $\mathbf{R}$, and the vector space $\mathbf{R}^3$ over $\mathbf{R}$. In these vector spaces, the vectors can be pictured by arrows in $\mathbf{R}^2$ and $\mathbf{R}^3$, respectively.

Volume. The volume of a cube of side length s is s^3. The volume of a cylinder of radius r and height h is $\pi r^2 h$. The volume of a tetrahedron or cone of base area A and height h is $Ah/3$. The volume of a sphere of radius r is $4\pi r^3/3$.

B
List of Bonuses

Cookie Jar Division	p. 2
Mediant Fractions	p. 3
Sum of an Arithmetic Progression	p. 5
Finding a Polynomial	p. 6
Curious Identities	p. 7
A Diophantine Equation	p. 9
Finding the Prime Factorization	p. 10
Ordered Digits in a Square	p. 11
The On-Line Encyclopedia of Integer Sequences	p. 12
Clock Magic	p. 13
Centigrade to Fahrenheit	p. 15
Sum of a Geometric Series	p. 16
Power Means	p. 18
Distributions, Partitions, and Schur's Estimate	p. 19
A Vector Proof	p. 20
A One-Triangle Proof	p. 21
Another Memorable Triangle	p. 22

B List of Bonuses

A Squarable Lune	p. 23
Covering With Unit Squares	p. 26
Calculation by Geometric Series	p. 27
Symmetry Groups	p. 27
Irrationality of $\sqrt{2}$	p. 29
Elementary Symmetric Polynomials	p. 31
Another Zigzag Path	p. 33
A Sum of Areas	p. 34
A Can of Minimum Surface Area	p. 35
The AM–GM Inequality	p. 37
Convex Functions	p. 38
What About the Second Hand?	p. 42
Sums to Any Number	p. 44
Inside or Outside?	p. 45
Lagrange's Interpolation Formula	p. 46
Integer Solutions	p. 47
An Algebraic Proof	p. 50
The Genus of a Graph	p. 52
The Problem for Hyperbolas and Parabolas	p. 54
Isometries of the Plane	p. 54
A Fundamental Parallelogram	p. 57
Tiling With Triangles	p. 58
A Family of Triangle Pairs	p. 61
The Wallace–Bolyai–Gerwien Theorem	p. 62
The Case of the Rotating Parallelogram	p. 64
Estimating the Harmonic Sum	p. 67
Another Quick Integral	p. 69
Probability that Two Numbers are Coprime	p. 70
Four Spherical Triangles	p. 71

B List of Bonuses

Fermat's Last Theorem	p. 72
Irrationality of π	p. 73
Average Number of Fixed Points of a Permutation	p. 75
Average Waiting Time	p. 76
A Two-Urn Problem	p. 77
Random Arcs on a Circle	p. 80
Average Number of Cycles in a Permutation	p. 81
Pell Equations	p. 84
Divisibility Rules	p. 85
The Folium of Descartes	p. 86
A Googol	p. 87
Ackermann's Function	p. 87
A Square Root Algorithm	p. 89
Powers of Fibonacci Numbers	p. 91
The Subcommittee Identity	p. 93
Fermat's (Little) Theorem	p. 96
Conway's LUX Method	p. 98
The Passage to Real Numbers	p. 100
An Instant Identity	p. 101
Packing 3-D Animals	p. 102
Linear Elastic Collisions	p. 104
The Correct Number of Orders	p. 106
Partitioning Space	p. 108
Other Large Integers	p. 109
Finite Derivative and Integral	p. 110
Arbitrary Line Sums	p. 111
Integer Averages	p. 112
Variations on the Games	p. 115
A General Drawing Strategy	p. 116

A Finite Field Method	p. 118
Brouwer's Fixed-Point Theorem	p. 121
Rational Generating Functions	p. 122
More Change for a Dollar	p. 124
A Generating Function for the Diagonal	p. 125
Non-Intersecting Transformations	p. 130
Lattice Point Vertices	p. 132
Dilworth's Lemma	p. 134
The Entropy Function	p. 137
Channel Capacity	p. 139
The Hat Problem	p. 142
The Catalan Numbers Modulo 2 and 3	p. 145
Flat Cyclotomic Polynomials	p. 148
A Tiling of the Hyperbolic Plane	p. 154
De Polignac's Formula	p. 155
Repeated Numbers in Pascal's Triangle	p. 157
Never a Prime	p. 159
Hilbert's Tenth Problem	p. 161
The Fundamental Theorem of Symmetric Polynomials	p. 164
Alcuin's Sequence	p. 167
The Probabilistic Method	p. 168
Dancing Links	p. 172
A Combinatorial Problem	p. 174
The Hoffman–Singleton Graph	p. 178
Asymmetric Graphs	p. 181

Bibliography

[1] G. Bachman. Flat cyclotomic polynomials of order three. *Bulletin of the London Mathematical Society*, 38:53–60, 2006.

[2] E. T. Bell. *Men of Mathematics*. Simon and Schuster, New York, 1937.

[3] M. Bernstein. The Hat Problem and Hamming Codes. *Focus*, 21(8):4–6, 2001.

[4] G. Cairns, M. McIntyre, and J. Strantzen. Geometric proofs of some recent results of Yang Lu. *Mathematics Magazine*, 66(4):263–265, 1993.

[5] M. Erickson and A. Vazzana. *Introduction to Number Theory*. Chapman & Hall/CRC Press, Boca Raton, Florida, 2008.

[6] M. J. Erickson. *Introduction to Combinatorics*. Wiley, New York, first edition, 1996.

[7] M. Gardner. *The Colossal Book of Mathematics: Classic Puzzles, Paradoxes, and Problems*. W. W. Norton & Company, New York, 2001.

[8] Liang-shin Hahn. *Complex Numbers and Geometry*. Mathematical Association of America, U.S.A., 1994.

[9] N. Hartsfield and G. Ringel. *Pearls in Graph Theory: A Comprehensive Introduction*. Academic Press, New York, second edition, 1994.

[10] A. M. Herzberg and M. Ram Murty. Sudoku squares and chromatic polynomials. *Notices of the American Mathematical Society*, 54(6):708–717, 2007.

[11] R. Honsberger. *Mathematical Gems I*. Mathematical Association of America, U.S.A., 1973.

[12] James P. Jones. Diophantine representation of the Fibonacci numbers. *Fibonacci Quarterly*, 13:84–88, 1975.

[13] K. Kedlaya, B. Poonen, and R. Vakil. *The William Lowell Putnam Mathematical Competition, Problems and Solutions 1985–2000*. Mathematical Association of America, Washington, D. C., 2002.

[14] T. Y. Lam and K. H. Leung. On the cyclotomic polynomial $\phi_{pq}(x)$. *American Mathematical Monthly*, 103(7):565–567, 1996.

[15] J. H. van Lint and R. M. Wilson. *A Course in Combinatorics*. Cambridge University Press, Cambridge, 1992.

[16] Y. V. Matiyasevich. *Hilbert's Tenth Problem*. MIT Press, Cambridge, MA, first edition, 1993.

[17] G. Pólya. *How to Solve It: A New Aspect of Mathematical Method*. Doubleday, New York, 1957.

[18] J. Rotman. *The Theory of Groups: An Introduction*. Allyn and Bacon, Boston, second edition, 1973.

[19] R. P. Stanley. *Enumerative Combinatorics*, volume 1. Cambridge University Press, New York, 1999.

[20] R. P. Stanley. *Enumerative Combinatorics*, volume 2. Cambridge University Press, New York, 1999.

[21] I. M. Yaglom. *Geometric Transformations I*. Mathematical Association of America, U.S.A., 1962.

[22] I. M. Yaglom. *Geometric Transformations II*. Mathematical Association of America, U.S.A., 1968.

Index

Ackermann's function, 87
Ackermann, Wilhelm, 88
Aha! Gotcha, vii
Aha! Insight, vii
Alcuin, 165
Alcuin's sequence, 165, 167
algebra, 1, 19, 84
algorithm
    depth-first search, 171
    gobbling, 80
alternating group, 28
AM–GM–HM inequalities, 18, 183
analysis, 46, 69, 84
analytic geometry, 86
animals, 101
annulus, 127
antiderivative, 68
arc length, 73
area, 19, 20, 26, 62, 68, 183
    of circle, 72
    of equilateral triangle, 133
    of quadrilateral, 19
    of triangle, 192
arithmetic, 1
    modular, 168
arithmetic mean, 18
arithmetic mean–geometric mean inequality, 37
arithmetic progression, 5, 159
asymptotic, 68
automorphism group, 152

Bachman, Gennady, 148
Banach's matchbox problem, 79
Banach, Stefan, 79
barycentric coordinates, 121

base-10 representation, 85
base-2 representation, 155
base-b representation, 85
Bell, E. T., 5
Berlekamp, E. R., 130
binary matrix, 170
binomial coefficient, 76, 94, 107, 154, 155, 183
binomial distribution, 136
binomial random variable, 135
binomial theorem, 166, 184
bits of information, 137
Boardman, Michael, 7
Boltzmann, Ludwig, 137
Bonaparte, Napoleon, 49
Brahmagupta, 84
Brahmasphuta-siddhanta, 84
Brouwer's fixed-point theorem, 121
Brouwer, Luitzen Egbertus Jan, 121
Burnside's lemma, 180

calculus, 20, 36
Callan, David, 111
capacity, 140
capacity function, 137, 139
Cassini's identity, 156, 161
Cassini, Giovanni Domenico, 156
Catalan number, 143
Catalan, Eugène Charles, 143
Cauchy, Augustin Louis, 39
celestial mechanics, 46
Celsius, Anders, 15
Centigrade, 14
centroid, 133, 192
channel
    binary symmetric, 139

capacity of, 140
channel capacity, 137, 139
characteristic polynomial, 163, 186
checkerboard, 102
chess, 124, 151
Chinese remainder theorem, 160, 184
circle, 23, 25, 33, 53, 56, 80, 184
    area of, 72, 183
    circumference of, 33, 72
circumcenter, 192
classical mechanics, 46
clock, 13, 41
coefficient of growth, 139
coin, 123
    biased, 138
    unbiased, 135
combination, 94, 184, 190
combinatorics, 93, 175
complete bipartite graph, 51
complete graph, 187
complete residue system, 118
complex number, 51
composite number, 85
computer science, 16
cone, 64
    volume of, 192
congruence, 158
conic sections, 184
conjugation, 58
conservation of momentum and energy, 104
constant of proportionality, 20
construction, straightedge and compass, 23
convergence of series, 17
convex function, 38, 142
Conway, John H., 98
counter, 110
covering system, 158
Csicsery, George Paul, 93
cube, 102
    symmetry group of, 27
    volume of, 192
curve, 130
cycle type, 180
cyclotomic polynomial, 146
    flat, 148
cylinder, 35
    volume of, 192

Damiano, Pedro, 151
Dancing Links, 172
data structure, 172
de Polignac's formula, 155
De Polignac, Alphonse, 155

decision procedure, 162
degree, 187
Delannoy path, 127
Delannoy, Henri, 127
derivative, 71, 185
    finite, 110
derivative operator, 74, 110
Descartes' rule of signs, 66, 185
Descartes, René, 86
determinant, 44, 133, 143, 156, 185
dice, 75
difference of squares, 85
digit, 11, 84, 108
Dilworth's Lemma, 134
Dilworth, R. P., 134
Diophantine equation, 9, 161
Diophantus, 9, 72
discriminant, 53
disk, unit, 70, 121
distribution, 19, 184
distribution problem, 106
divisibility rule, 85
divisor, 12
draw game, 115
Dudeney, Henry Ernst, 62
"duplication method", 4

elementary symmetric polynomial, 30
ellipse, 53, 184
elliptic curve, 72
entropy function, 136
equation
    Diophantine, 161
    quadratic, 191
Erdős–Selfridge theorem, 116
Erdős, Paul, 93, 112, 116, 159, 168
error-correcting code, 140
Euclidean plane, 54, 58
Euclidean space, 54
Euler's formula, 31, 185
Euler's function, 146, 185
Euler, Leonhard, 69
Even Steven, 113
exact cover, 170
exact cover problem, 170
existence argument, 168
expectation, 75, 81
expected value, 74
exponential function, 139

factorial, 84, 190
Fahrenheit, 14
Fahrenheit, Daniel, 15

fair division, 120
Falco, Marsha, 172
Fano, Gino, 151
Farey sequences, 3
Fermat's (little) theorem, 95, 96, 112, 186
Fermat's Last Theorem, 72
Fermat, Pierre de, 72
Fibonacci number, 12, 122, 156, 161, 186
Fibonacci sequence, 12, 186
The Fibonacci Quarterly, 160
field, 118, 186
    finite, 118, 187
finite difference calculus, 6
Fishburn, Peter, 134
fixed point, 54, 75, 121, 180
Folium of Descartes, 86
fraction, 2
    mediant, 3
function, 185
    convex, 38
    doubling, 87
    exponential, 87
    linear, 38
    primitive recursive, 88
    rational, 65, 122, 158
    recursive, 88
    sine, 69
    tower, 88
functional analysis, 79
fundamental theorem of arithmetic, 44, 187
fundamental theorem of symmetric polynomials, 164

game, 113, 138
    Even Steven, 113
    Oddball, 113
    SET, 172
    taking, 116
    tic-tac-toe, 102, 115
game theory, 16, 102
Gangolli, Anil, 11
Gardner, Martin, vii, 42
Garns, Howard, 169
Gauss, Carl Friedrich, 5
generating function, 122, 123, 125, 158, 165, 187
    algebraic, 126
    rational, 126, 167
genus, 52
geometric construction, 32
geometric mean, 18
geometry, 1, 84
    analytic, 86

Gilbert, E. N., 130
Ginzburg, A., 112
glide-reflection, 54
globe, 71
Göbel, Frits, 25
golden ratio, 22, 28, 130, 187
Golomb, Solomon, 102
googol, 87
Gould, Wayne, 169
graph, 9, 178, 187
    asymmetric, 181
    complete, 167
    cycle of, 175
    girth of, 175
    Petersen, 178
    regular, 176
graph theory, 69, 93
graph, directed, 185
Gray code, 110
Gray, Frank, 110
greatest common divisor, 188
group, 59, 148, 180, 188
    abelian, 100, 188
    alternating, 28
    automorphism, 178
    general linear, 151
    matrix, 149
    of symmetries, 154
    presentation, 148
    special linear, 153
    symmetric, 28
    triangle, 154
group action, 188
group presentation, 59

Hahn, Liang-shin, 51
Hall's Marriage Theorem, 175
Hall, Philip, 175
Harary, Frank, 102
harmonic mean, 18
harmonic series, 67
harmonic sum, 67, 81
Hartung, Paul, 160
Heron's formula, 60, 192
hexagon, 50
Hilbert's Tenth Problem, 161, 162
Hilbert's Third Problem, 63
Hilbert, David, 63, 162
    23 problems, 162
Hippocrates of Chios, 23
Hoffman, Alan, 178
Honsberger, Ross, 107
Horace, vii

hyperbola, 9, 54, 61, 184
hyperbolic plane, 59, 154
hyperplane, 108

identity operator, 74
incenter, 192
inequalities
    AM–GM–HM, 183
inequality, 2, 23
    AM–GM, 37
    AM–GM–HM, 18
    Jensen's, 38
    mediant fractions, 3
    triangle, 192
Information Theory, 136, 139
integer triangles, 2
integral, 72, 188
    definite, 68
    finite, 110
integral operator, 110
intermediate value theorem, 188
interval, 134
irrational number, 28
isometry, 54
isoperimetric inequality, 36

Jensen's inequality, 38, 142
Joint Mathematics Meetings, 140
Jones, James P., 160
Jordan curve theorem, 45, 188

Kemnitz, Arnfried, 48
Knuth, Donald, 11, 170, 172
Kummer surface, 49
Kummer, Ernst Eduard, 49, 155

Lagrange's interpolation formula, 46
Lagrange's theorem, 95, 188
Lagrange, Joseph-Louis, 46
lattice point, 132, 188
lattice theory, 134
Laurent series, 127
Laurent, Pierre Alphonse, 127
law of cosines, 47, 64, 188
law of sines, 189
least common multiple, 189
l'Hôpital's rule, 141, 142, 189
line, 32, 118
linear elastic collision, 104
linear recurrence relation, 167
lines, perpendicular, 53
linked lists, 172
locus, 53
Lucas number, 91, 123

Lucas, Edouard, 162
lune, 23
LUX method, 98

magic square, 96
Martin, Gary, 100
matches, 79
mathematical induction, 39
mathematical notation, 69
Matiyasevich, Yuri, 162
matrix, 44, 110, 143, 153
    binary, 170
    determinant of, 44, 143
    eigenvalue of, 177
    eigenvector of, 177
    invertible, 94, 151
    nonsingular, 174
    orthogonal, 56
    permutation, 111
    rotation, 50
    symmetric, 176
matrix group, 149
McDaniel, Wayne, 157
mean
    arithmetic, 18, 183
    geometric, 18, 183
    harmonic, 18, 183
median, 192
mediant fraction, 3
Michigan, University of, 102
Möbius, 147
Möbius inversion, 147
modulo 2, 46
multinomial coefficient, 189
multinomial theorem, 112, 189

Napoleon's theorem, 49
necklace, 96
Nelson, Roger B., 5
Newton's identities, 32
Newton, Isaac, 61
N is a Number: A Portrait of Paul Erdős (film), 93
Niven, Ivan, 73
non-self-intersecting, 130
number
    algebraic, 190
    Catalan, 143
    complex, 31, 51, 146
    composite, 190
    Fibonacci, 12, 122, 156, 161, 186
    irrational, 9, 28, 29, 189
    Lucas, 123

prime, 10, 159, 190
rational, 9, 99, 189
sequence, 186
square, 11, 12, 83, 132
transcendental, 190
triangular, 83
number theory, 46, 69, 84, 93

octahedron, 27
Oddball, 113
odometer, 109
operator
    derivative, 74, 110
    identity, 74
    integral, 110
orbit, 179
Oresme, Nicole, 67
orthocenter, 192

parabola, 54, 61, 184
parallelogram, 20, 23, 57, 62, 64, 91, 130
    area of, 183
    Varignon, 20
parallelogram law, 57
parameterization, 10
parity, 45, 115
partial fractions, 126, 166
partial order, 134
particle–particle collisions, 103
partition, 19
Pascal's identity, 190
Pascal's triangle, 12, 93, 157, 190
path, 143
    shortest, 32
Pell equation, 84
Pell, John, 84
pentagon, 129, 130
pentagram, 129
period, 167
permutation, 75, 100, 108, 180
    cycle notation, 81
    cycle type of, 81
    random, 81
Perrin's sequence, 162
Perrin, R., 162
Petersen, Julius, 178
π, 70, 72
pigeonhole principle, 134, 190
plane, 107
point–slope formula, 46
points
    at unit distance, 131
    in $\mathbf{R}^d$, 131

pole, 158
Pólya, George, 180
polygon, 129
    self-intersecting, 130
polynomial, 6, 46, 161, 190
    cyclotomic, 146
    elementary symmetric, 30, 31
    flat, 148
    irreducible, 118
    monic, 146
    quintic, 65
polyominoes, 102
power mean, 18
power series, 69, 111, 187
power set, 190
prime number, 10, 12, 159, 190
principle of inclusion and exclusion, 80, 191
probabilistic method, 168
probability, 81, 82
projection, 70
proof by contradiction, 73
The Proof (film), 72
Ptolemy's theorem, 48, 191
Puiseux's theorem, 126
Puiseux, Victor, 126
Pythagorean theorem, 20, 24, 34, 191

quadratic equation, 191
quadratic formula, 22, 126, 132, 191
quadratic residue tournament, 168
quadrilateral, 19, 130
    cyclic, 191

random variable, 75
    binomial, 135
random walk, 79
rational root theorem, 191
rational solution, 9
real numbers, 100
rectangle, 36, 62, 63, 69
recurrence relation, 83, 122, 145
    linear, 90, 167
    palindromic, 167
recursion, 88
reflection, 54
relativity theory, 93
Riordan, John, 92
rotation, 54
rotation matrix, 50
rotation operator, 57

salt water, 3
scalar, 192

Schoenberg, Isaac Jacob, 133
Schur's estimate, 19
Schur, Issai, 19
Selfridge, John, 116
semiperimeter, 192
sequence, 122
    D-finite, 127
    Alcuin's, 165, 167
    Fibonacci, 12, 156
    Perrin's, 162
series, 191
    arithmetic, 191
    geometric, 16, 191
SET game, 172
Shannon's First Theorem, 137
Shannon's Second Theorem, 140
Shannon, Claude, 137
simplex, 131
Simpson's Paradox, 3
Sinden, F. W., 130
sine curve, 69
Singh, Simon, 72
Singleton, Robert, 178
Sirotta, Milton, 87
Sloane, Neil J. A., 12
solid of revolution, 63
spectral theorem, 177
Sperner's lemma, 120
Sperner, Emanuel, 120
sphere, 52
    unit, 70
    volume of, 192
spline method, 133
square, 8, 25, 62, 83
    area of, 183
    unit, 26
square root, 8, 89
Stanley, Richard, 111
Stevens, W. L., 80
Stirling's approximation, 135, 191
subcommittee identity, 93
subset, 11
Sudoku, 169
sum
    harmonic, 67
    infinite, 69
surface, 52
    irrational, 49
    Kummer, 49
    quartic, 49
    tertrahedroid, 49
surface area, 35
symmetric group, 28

symmetric point, 32
symmetry, 35, 59, 68, 154, 178, 181
    circular, 96
symmetry group, 27
system of distinct representatives, 175
Szekeres, George, 93

tangent line, 53
Taniyama–Shimura conjecture, 72
tessellation, 62
tetrahedroid, 49
tetrahedron, 27, 28, 63
    regular, 131
    volume of, 192
The On-Line Encyclopedia of Integer Sequences, 12, 125
tic-tac-toe, 102, 115
tie series, 78
tiling, 58, 154
torus, 51
tournament, 167, 192
Tower of Hanoi puzzle, 143
train, 15
transformation
    isometry, 54
    orthogonal linear, 55
    similarity, 131
translation, 54
transvection, 153
transversal, 174
triangle, 2, 20, 23, 33, 37, 58, 60, 62, 67, 83, 120, 154, 165, 192
    altitude of, 192
    area of, 60, 183
    circumcircle of, 48
    equilateral, 26, 46, 49, 62, 131
    inradius of, 34, 60
    isosceles, 61
    perimeter of, 60
    right, 20
    semiperimeter of, 60
    spherical, 71
triangle group, 154
triangle inequality, 23, 165, 192
triangle, "centers" of, 192
triangular wave function, 103
trigonometry, 29, 84
Twin Prime Conjecture, 155

Van Schooten, Franz, 61
Varignon parallelogram, 20
Varignon, Pierre, 20
vector, 20, 192

dot product, 55
vector space, 192
    basis for, 94, 174
Venn diagram, 191
volume, 27, 35, 63, 192
    of cone, 64
    of regular simplex, 133
    of tetrahedron, 27
von Neumann, John, 16

waiting time, 76
Wallace–Bolyai–Gerwien theorem, 62
wallpaper pattern, 50
Wiles, Andrew, 72
William Lowell Putnam Mathematical Competition, 99, 141

Yaglom, I. M., 33

zigzag sequence, 167
Ziv, A., 112

About the Author

Martin Erickson is Professor of Mathematics at Truman State University. He received his PhD at University of Michigan in 1987. He has authored three mathematics textbooks: *Introduction to Number Theory* (with Anthony Vazzana; Chapman & Hall), *Introduction to Combinatorics* (Wiley), and *Principles of Mathematical Problem Solving* (with Joe Flowers; Prentice–Hall). Professor Erickson is a member of both the American Mathematical Society and the Mathematical Association of America.

LIBRARY
Lyndon State College
Lyndonville, VT 05851